MEIGS LINE

Rangers rediscover a two-century-old disputed
boundary between the U.S. and
Cherokee Nation

MEIGS LINE

Rangers rediscover a two-century-old disputed boundary between the U.S. and Cherokee Nation

Dwight McCarter
Joe Kelley

Grateful Steps
Asheville, North Carolina

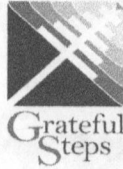

Grateful
Steps

159 South Lexington Avenue
Asheville, North Carolina 28801

McCarter, Dwight and Kelley, Joe
Meigs Line
Rangers rediscover a two-century-old disputed boundary
between the U.S. and Cherokee Nation

ISBN 978-1-935130-91-8 Paperback

Cover design by Sundara Fawn and Dwight McCarter
All photographs by Dwight McCarter
and Joe Kelley unless otherwise specified.

Printed at Lightning Source

6   5   4   3   2

Library of Congress Cataloging-in-Publication Data

McCarter, Dwight, 1945-
  Meigs line : rangers rediscover a two-century-old disputed boundary between
the U.S. and Cherokee nation / Dwight McCarter, Joe Kelley.
     p. cm.
  ISBN 978-1-935130-91-9 (paperback : alk. paper)
  1. McCarter, Dwight, 1945---Fiction. 2. Kelley, Joe, 1939---Fiction. 3. Park
rangers--Fiction. 4. Great Smoky Mountains (N.C. and Tenn.)--Fiction. 5.
Surveyors--Fiction. 6. Surveying--Fiction. I. Kelley, Joe, 1939- II. Title.
  PS3613.C34565M45 2009
  813'.6--dc22

                                              2009007284

www.gratefulsteps.org

for Julia Kelley
May 1, 1939 – January 13, 2007

for Polly Davis
Newberry, SC

This book is dedicated to Polly Davis of Newberry, South Carolina. She was a teacher for thirty years in the South Carolina system. Polly keenly appreciates the ever-increasing value of this historical book and guided and aided me in its completion. Polly loves the Lord. She loves and cares for her neighbors. Polly loves and cares for God's little animals. But best of all, Polly loves me. And I definitely love her.

– Dwight.

PREFACE

On a foggy day in July of 1970, two Great Smoky rangers parked at the Spruce Fir Nature Trail near Mount Collins on the Clingmans Dome Road. John O. Morrell had just retired after decades of service to the park. Joe Kelley's career was just a few years old. Their plan was to ascend Mount Collins and find Meigs Post. John Morrell, along with Fire Control Aid George Lamon, had been to the site in the 1950s when they reset the post using a concrete highway-type marker topped with a brass button on which was inscribed

U.S.D.I. MEIGS POST.

"At least we're above the rain," John said as they walked through the fog along the Appalachian Trail toward Mount Collins. The nettle weeds were thick and wet and stung the skin unmercifully on this damp, misty day. They had just begun to descend the west side of Collins when John said, "Well, we've walked past it. It's somewhere behind us."

They had retraced their steps about 100 yards when John saw a large moss-covered axe blaze mark on a large balsam tree on the east side of the trail, which he recognized as one of the historic witness trees he had seen before. "It should be somewhere on our left—very close by!" After more nettles, wet sleeves and stinging hands, finally, there it was—Meigs Post!

Great Smokies Park Ranger Joe Kelley, Meigs Post, July, 1970

John soon spotted six additional witness trees, referenced in Return Jonathan Meigs' 1802 journal.

"That was an overwhelming moment for me!" Joe said later. "To touch those trees that the actual hands of earlier Americans had blazed so long ago!"

Recently, John had told Joe of the Ben Hawkins survey line of S76°E—how it began in 1797 at Kingston, Tennessee, and was, at that time, the southern boundary of the United States, interfacing the territory of the Cherokee Nation. He shared with him how Pickens had finished Hawkins' work, and then Butler reset the line. He emphasized that Meigs performed the definitive survey in 1802. Meigs Post, set atop Mount Collins within eight miles of the northwest end of Meigs Line, had been lost, found and lost again on numerous occasions.

Morrell told Joe about the section of the line on Chestnut Top—now part of the boundary of the Great Smoky Mountains National Park—that came very close behind the Tremont Ranger Station where Joe and his family lived at that time. He spoke of the details, how the line had been at the root of so many boundary disputes of folks living in the Little River Valley as he tried to work over tough and bitter years with them on the park's land acquisition program managed by the Great Smoky Mountains Park Commission. Joe became obsessed with the question of whether this line had ever truly been "put on the ground" between the high points and ridge tops.

***

Dwight McCarter, a best-selling author and ranger with national prestige for his tracking skills and wilderness savvy, learned of Joe's passion to explore the historic survey. In the mid 1970s, Dwight walked the historic lines. He began at Fort Southwest Point in Kingston, Tennessee, followed Hawkins Line and Pickens Line S76°E, then Butler Line and, finally, the ultimate Meigs Line. When not in the wilderness, he drove the rural roads through towns and woods and—with trained recognition of historic blazes—rediscovered evidence of the original surveys, using compass and camera.

Dwight spent months poring over ancient maps, studying the portion of Meigs' journal still in existence, researching in libraries, contacting surviving relatives of Return Jonathan Meigs and talking with those expert in the history of the region. Then to share the amazing story, he wrote it all down, transcribed it to an electronic format and submitted it for publication.

When Sweet Julia, Joe's wife, died, Dwight chose to make the book a memorial to her, as well as a tribute to his beloved Polly, and invited Joe to join him as a contributor. The information was expanded to include more ranger wilderness lore and details of

John Morrell's first-hand knowledge of the survey lines and their relationship to the Smoky Park and the region.

***

The points and references of the surveys listed in this work are real. Every event experienced by the characters Vinn and Dwight in this work are based on real happenings during the careers of Rangers Dwight McCarter and Joe Kelley. They represent smiles, tears, pains and joys experienced by both. People's names and timing of events were modified occasionally to enhance readability.

INTRODUCTION

High atop the Great Smokies, on Mount Collins, there is a point known as Meigs Post, the most important survey point used in the establishment of the Great Smoky Mountains National Park.

The story of Meigs Post begins when our nation was brand new. Our country had obtained her freedom from the British and was searching for her boundaries. This post marks the spot that was once the common corner between the states of Tennessee, North Carolina and the Cherokee Nation as agreed upon in the Treaty of Holston, ratified by Congress on July 2, 1791. Both parties—the treaty commissioners and Cherokee involved—agreed to the location of elevation 6003 feet on the North Carolina State Line as one endpoint, and the other, where the Clinch joins the Tennessee River. This treaty line, to be surveyed in 1792, was to begin at Fort Southwest Point, now Kingston, Tennessee, and extend on a compass course S76°E. In 1792 Campbell, McKee and McClung ran an exploratory line. So many white settlers were found south of the line, it was abandoned.

A complete survey was not accomplished until some five years later; the federal government commissioned a better survey of the treaty line that began at Fort Southwest Point and ended at the Great Smoky Mountains at the present Mount Collins and,

Gen. D. Smith Map displaying Hawkins Line, showing 12 mountains between Fort Southwest Point and top of the Great Iron Mountians (Smokies)

because of additional cessations, was extended to the South Carolina line near Tryon, North Carolina.

In March 1797, President Adams commissioned Colonel Benjamin Hawkins, General Andrew Pickens and General James Winchester to survey the line. They were to meet with twenty Cherokee representatives at Tellico Blockhouse on the Little Tennessee River. Winchester never appeared at the meeting.

The survey was begun on April 5, 1797 with Hawkins, Pickens and Timothy Barnard. A close friend of Colonel Hawkins, Barnard was a trader and surveyor from the Flint River near Fort Fidius, South Carolina. He was on the survey team as the compass man. The survey started one thousand yards northeast of Fort Southwest Point at the blockhouse spring and was on azimuth S76°E.

It took eleven weeks to run the sixty miles of the Hawkins Line. Hawkins ended the survey at Mount Collins on July 1, 1797. The first survey marker at the 6003-foot site was most likely erected by Colonel Benjamin Hawkins in 1797. We can assume that Hawkins erected some rough-hewn type of marker.

The project was abandoned when he had been ordered back to Knoxville to investigate treason charges against a high-ranking official. In October 1797, furious settlers ran their own line and held to their land claims south of the boundary. In November 1797, Hawkins returned, this time with federal troops, and evicted settlers from south of the line. In the October 1798 U.S. treaty with the Cherokee, the Indians ceded land from Chilhowee Mountain to the Little Tennessee River. The Hawkins Line from a point on Chilhowee Mountain to the Smoky Mountains remained unchanged. But in that same year, the Hawkins Line from Kingston to Chilhowee Mountain was made moot, a new line established to be defined by the Little Tennessee River. This survey generated many disputes. There was some question as to the survey line S76°E actually being "put on the ground" between Blanket Mountain and Mount Collins. Many "quit claim" deeds had to be obtained by the park from private landowners living in what is now the Elkmont area and the Upper East Prong of Little River.

Colonel Thomas Butler conducted a survey in 1799, designed to appease the white settlers outside the southern line. He over corrected and infuriated the Cherokee, many now finding themselves in white territory subject to eviction. Colonel Return J. Meigs, the United States Indian Commissioner, was commissioned to settle the matter with a survey line to appease all parties.

Meigs referred to the Hawkins' survey marker as a "point of view." This probably was in some area of cut forest that could be viewed from afar. In 1802, Meigs Post was placed—made of wood and 6' x 8" x 8" in size. Meigs Post was the beginning corner of the two largest Tennessee tracts of land acquired for the park—Little River Lumber Company's 56,000-acre tract that extended westward to Thunderhead Mountain and Defeat Ridge and the Champion Fiber Company's 39,000-acre tract that extended northeastward to Old Black and the Pinnacle Lead.

In the year 1790, the year before the Treaty of Holston was signed and a boundary line agreed upon, President George Washington had appointed William Blount to be the governor of the New Southwest Territories. In 1792 a blockhouse was built just below the present town of Kingston. The blockhouse was erected a very short distance from the only pure water source for the area. The spring was hidden under large flagstones, protecting the access inside the building from attackers. The occupants of the blockhouse could shoot from eight different points, making it impossible for enemy disruption of the water source. General John Sevier was in charge. Immediately, the work on Fort Southwest Point began, one thousand yards to the southwest.

In 1793 a small attachment of federal troops was established there, at Fort Southwest Point. In 1794 the federal troops began providing armed escorts for travelers between the blockhouse and Nashville. On June 1, 1796, Tennessee became a state. Fort Southwest Point was completed in 1797. The fort had a commanding view from the top of a hill, overlooking a bend of the Clinch River. Four hundred federal troops were stationed there, commanded by Colonel David Henley.

Restored Fort Southwest Point Circa 2009
Top photograph: Exterior view. Bottom photograph: Interior view

HAWKINS LINE

I first met Vinn Garoon in the spring of 1964. The old ranger was a likable sort, getting along with friend and foe alike. Old Vinn began as a young ranger in the Great Smoky Mountains National Park a year after the park was formed in 1934. Back then, he towered six foot seven inches and weighed a hefty 260 pounds. When he walked, his #15 boots sank into the earth.

Just discharged from the military, I got a job as a ranger in the Smokies. Vinn Garoon and I became fast friends, he in the autumn of his career as my career was just beginning. He always took time to give me pointers about the park and its visitors. While working on any projects with Vinn, his favorite subject would always come up—finding the Meigs Line.

Vinn Garoon, now in his mid-sixties, had some common ailments that attack us all in the later years. A mild heart ailment had put him on a strict diet, leaving him a shell of the man that he once was. But Vinn was as hard as nails. He had been tested in the demanding jobs that come with being a ranger. He could not remember the number of forest fires, rescues, floods and police searches he had been on.

***

As the time for retirement approached, old Vinn talked of his future trip to find a lost survey line commissioned by Thomas Jefferson. He showed me how the early surveyors had used paces and chaining to measure distance from one point to another.

Years earlier, as a "buck ranger," a brand new rookie, I had practiced with old Vinn to learn my pace. Vinn would measure 66 feet and I would walk it three times, counting the number of times my right foot hit the ground. The average number of times that my right foot hit the ground in 66 feet was 13. Vinn's hit the ground 12 times in 66 feet because his stride was longer than mine. Sixty-six feet equals one chain, and eighty chains equals one mile.

Vinn practiced diligently to prepare for his trip, using an old Jacobs staff compass, which looked like it was used by the 1802 surveyors themselves.

On boundary patrol, Vinn had shown me old axemarks on ancient oaks. These marks on a tree followed legal parameters. Some of these he had made himself when the park boundary was being surveyed. One axemark meant approaching a corner, and three axemarks meant changing direction.

I felt extremely lucky to get to work with Vinn. I learned things that I could never acquire through school or watching an official training film at the Great Smoky Mountains National Park.

This man had been there in the beginning of the park. He showed me photos of his first ranger truck—a 1933 Ford in which only two people could ride. It had a gearshift in the floor, and the starter was in the floor, as well. I told Vinn it had been common practice for the rangers to stop and give my dad and me a ride when we fished high up the valleys in the Smokies. I grew up half a mile from the park boundary.

"Come over here, lad!" Vinn said. "I have saved this old radio I used until the 1950s. Pick it up. It weighs 60 pounds. The batteries alone weigh ten pounds each. The radio was outdated and due to be thrown in the salvage. I was able to keep it as a backup in case the newer radios failed. To get a reception, I would always have to throw this antenna up over a primitive telephone wire.

"Personally, lad, I don't like the new-fangled radios. If you want to talk to Luftee, you have to turn the knob to channel C, and you still can't get Cataloochee. The same happens on channels A, B, and D. I do remember being one of the rangers that picked you up while you were high on Tremont, fishing. Did you ever get a pair of shoes? Every time you jumped in the back of the truck, you were barefoot and looked like you needed a good worming. Well, we no longer can give rides 'cause some feller fell out of the truck and sued us—such a shame. For we generated a lot of trust from locals, and in the process we even learned a lot about illegal activity. We lost our contact with people who had formed networks to respond immediately to fires, drownings and lost people."

Vinn spotted an old white insulator high up in the tree with thick copper wire running through it.

"These wires go to every fire tower and ranger station in the park. There are 1200 miles of wire in the park. This wire went to Rich Mountain Tower and my favorite, Blanket Mountain Tower. I worked that tower a lot, lad, and in my spare time, looked for markings on rocks and trees telling of the old survey lines' routes.

"I found Meigs Line came right over a beautiful cold spring that flowed from under a large flat rock. On top of this rock were names of different people some hard to read. And, lad, that is the only spring within one hour's hiking distance, so Meigs had to drink there. His party must have camped there. This spring is ice cold and is on the Marks Cove side of Blanket. I also found the big boulder outcrop near the tower where the survey party placed a bright red blanket to sight their compass on. Come on, lad, I'll buy you lunch. This is Charlie Ogle's store. Come on in. We'll have two thick bologna sandwiches with mayo, lettuce and pickle."

Vinn never asked if that's what I wanted.

"Here, lad, eat this. We'll put some meat on you, yet."

"Vinn, why do you have all those green paint pens in your dash?"

"Well, lad, when you've been here long as I have, you're allowed three wrecks a year. Those pens are for painting over minor scratches that don't rise to a real wreck."

***

The next morning I could hear Vinn calling someone over the radio.

"410! This is 410! 410! This is 410."

It occurred to me Vinn was calling himself. I answered him, "410! This is 412! Can I help you sir?"

"Yeah, where you been, Dwight? I been calling you."

"Sorry, sir, my volume was down."

"412! Come into the office! I got a job for you."

Wow! My first assignment. My chest swelled up. What could it be?

Vinn was waiting. "Lad, you see that there box in the desk? You take them leftover biscuits and gravy up to the top of the mountain at the gap. You give them to Oscar, my raven friend. He'll be waiting on the Appalachian Trail sign."

"How will I know it's Oscar, sir?"

"You'll know, son. Now get going. Oscar gets cranky when he misses a meal."

Arriving at the top, I saw ravens everywhere but one very fat raven came swooping down to land on the sign. Oscar scowled at me. "Auk, Auk," as if to say, "Where you been, lad?"

I don't know how Vinn knew Oscar was male. But he seemed to know all these things naturally. Oscar was never wasteful, and after his fill, he left the remainder for the less fortunate ravens.

I watched with amazement as Oscar tried to get airborne. Finally, he got a good run-n-go into the wind and banked sharply down the valley into North Carolina.

***

Vinn and Julia lived at Tremont Ranger Station during his late career. This station is very close to S76°E line. I was able to visit with him before he left on his venture. We spread maps out on a nearby school gymnasium floor and studied the area he would soon be walking. Both Vinn and I had years of experience in reading maps, a ranger skill used in every aspect of managing a major park.

Several days later Vinn told me his retirement date had been approved. The final date would be June 12, 1968.

"Dwight, this coming week on Monday, I will begin using my remaining leave. I will not return until my retirement party on June 12th. I plan on using my time hiking the original Hawkins, Pickens, Butler and Meigs Lines. My sweet Julia will drop me at Fort Southwest Point on Monday. I will see you on the 12th of June."

With that, Vinn was absent from the normal routine, and I busied myself with normal duties.

***

Julia had prepared me well with food, ample changes of clothes and cold weather gear. She dropped me off at 8 a.m. on Monday, April 7, 1968 near Fort Southwest Point, Kingston, Tennessee. With assurances to her that I would call her often, I embarked on my vast adventure. With my pack on my back and the old Jacobs staff compass, I climbed the hill to the old fort. I dug out my old Pentex 35mm and shot a view.

After entering the main gate, I was pleasantly surprised to find a handsome young man also photographing the area. He told me he was nine years old and that he was a boy scout visiting the fort with his troop. From over the original stone marker in the fort, I shot another picture with the young man there.

Restored Fort Southwest Point
Interior view showing young scout in foreground

I set the old staff compass directly over the spring at the Fort Southwest Point Blockhouse, three-quarters of a mile upstream from the junction of the Clinch River and the Tennessee River. I turned the setting to S76°E and walked off into history.

Restored Fort Southwest Point
Exterior view

WAYFARING STRANGER

As I walked, I sang the old Christian song "Working the Road to Glory." It felt great to be out doing things I liked. The staff compass pointed me to a distant ridge. The closer I got, the more clearly I could make out details. A large oak came into view. Arriving at Ridgecrest Drive, I stood in the road and viewed the large oak. It definitely had clear axemarks that I could distinguish as having been left long ago. The old blazes had healed over after many years of tree growth. The original marks were made with a person standing, swinging a broadaxe chest-high on the tree. Two cuts had originally been made the width of the broadaxe. I marveled at seeing something clearly from the eighteenth century. The mark doesn't grow up as the tree grows taller; rather, the axemarks expand more laterally than vertically—so they will still be at the same level, even centuries later.

The survey group had included 20 Cherokee axemen—the individuals who actually placed the blazes. Three Cherokee chiefs were with Hawkins, plus a military escort consisting of 60 troops.

Colonel Benjamin Hawkins
Photograph courtesy of North Carolina Office of Archives and History
Raleigh, North Carolina

The compass pointed me across the road and brought me to another road, a distance away. My map identified it as James Ferry Road. Standing along the road, I attracted many second looks from occupants in vehicles passing by. The compass pointed me toward a large silo among many cedar trees. *I recall reading in my Bible that someone paused to rest under a juniper tree. Well, if it's good enough for him, it's good enough for me.* The warmth of the sun shone upon me and I drifted off to sleep.

In my pleasant dreams, I could feel the sweet breath of Julia. I lifted one eyelid. It wasn't sweet Julia. The biggest cow I had ever seen was face to face with me. We both scared each other and ran away in opposite directions.

Gathering my composure, I put the staff compass to the ground. It pointed me toward more cows, bales of hay in a barn, and barking dogs. The dogs quickly discovered me. Some just barked, some liked to sniff, and some wagged their tails. I got to meet all the dogs in that nice community. The dogs were even kind enough to alert other dogs ahead so they could also meet me.

The compass led me past junk cars. *I think I once owned one like that one. It was so long ago.*

I came past frogs in a pond, more dogs, honeysuckle and ever-present briars. In a big sage field I passed under a huge power line, probably from the Kingston Coal Power Plant. After crossing another country road, I found myself in the middle of a golf course amidst patrons at play. Exiting the golf course, I crossed Paint Rock Ferry Road. A kind gentleman sitting on his porch offered a cold drink. I sat on the stoop and conversed with him.

"My dog had told me you was coming," he said. "Ol' Paint wanted to go meet you, but he figgered you'd be here soon enough."

"Yes, sir. It seems I've met all the dogs in your neighborhood." I thanked him and continued following my staff compass. It led me to the top of a big flat piney ridge. There on the ridge was an old, gnarled oak with the tell-tale axemarkings. It seemed a pleasant place to stay. It would be my camp for the night. The evening was cool. As I ate supper, I could see off in the distance a large lake. My map said it was Watts Bar Lake. Late into the night, the barred owls sang their song, "Who Cooks For You?" I never realized how many sore muscles one could have.

The next morning I packed my stuff as the sun peeked over the Smokies, still far away. I headed down a steep hill and left the pine ridge behind. A valley appeared in front of me with green fields and very few homes. The compass pointed directly toward a large brick church.

The Cedar Grove Baptist Church came into view as I crossed Johnson Valley Road. The line ran between two large maples as if it was a gunsight. Twenty feet to the right of the church was a big oak, right on the line. Further along were two large cedar trees dead center of the line. It was time to give Mr. Jacobs staff a rest. I sat on the steps of the beautiful church and studied my surroundings. There was a cemetery that had been obviously well cared for. The church grounds were immaculate.

I ascended a valley that had quite a few homes. Hawkins Line came near an area where a mobile home has just been moved away. There against the bank was a little 125 Honda trail bike. *If I only had one of those, I could be singing "Riding the Road to Glory."*

I topped out at a deep gap on Chestnut Ridge. *Surely this deep gap would have been mentioned in the original survey. Too bad a lot of the records were lost, when much of Washington burned during the War of 1812.*

Down the other side, I descended from the ridge to a three-way road junction. At the intersection, an elderly couple came out to look me over. The man said, "I live here."

"Yes, sir," I replied.

"I live here," he came again.

Not wanting to tread any closer to their property, I waved goodbye to the folks and continued working my way to glory.

From the three-way, I followed the compass down a ridge to Buck Creek. From there, the line took me on a climb up Dug Ridge—elevation 1120 feet. I could not locate any marked trees on the ridge. *Perhaps logging activity removed those old treasures.*

***

The compass then led me down into New Midway Valley. The valley has an abundance of old farms and pretty fields. I came upon a tree with a homemade poster "Lost dog female $100.00 reward." As I crossed Greasy Creek at New Midway Road, I had already progressed six miles from the old fort.

From the lost dog poster, I climbed up Calloway Hollow to the top of a small mountain. There at the end of a dead end road I came to a big oak—blazed and on line!

I rested at the foot of the historic tree and enjoyed pears that I had bought at the country store nearby.

Not wanting to spend another night on a piney ridge, I stopped at a deep little hollow. I made a quiet camp under a large white pine. Lots of different birds called to each other as darkness arrived. I could hear them late into the evening, adjusting their position to be closer to each other. I was able to make out the calls of cardinals, the Carolina wren and the hermit woodthrush.

I thought often of my sweet wife. *She understands that this walk is something that brings joy to me. On the joy side, it is nice to be free of everyday chores and see new places and faces. But on the down side, I find I must wash my dirty socks every evening and let them dry under my tarp. I will look for a Laundromat at Lenoir City.* I slept well. The soft pine needles made a good bed.

The next day I left the quiet hollow and climbed a big fat mountain. The top was so big and wide, I would have been confused without my trusty staff compass. Dropping off the big fat mountain, I found no trees with axemarks. But as I came to some splendid green mountain pasture, I could see a huge sycamore tree in the hollow. Taking my staff compass. I brought it to bear on S76°E. I looked down the sights and was delighted to see the sycamore tree was dead center. Hurriedly, I went to the tree. This was like candy to a kid for me—the solving of a long-ago mystery—a line that was debated, litigated and literally fought over.

The tree had axemarks and names on it I could not read. *I will return with chalk later. The chalk will tell the story. I will use white chalk to highlight the edges of the knife marks as I travel about through the forest. I will leave no trace of my comings or goings. The first rain will wash away my chalk markings and my footprints.*

As I left the sycamore tree, I remembered a Bible reference to a man who climbed a sycamore tree to view Jesus.

> *Zaccheus was a wee little man*
> *A wee little man was he.*
> *He climbed up in a sycamore tree . . .*

How does the rest of that go? Sweet Julia will know. I'll ask her.

Well, I had to sing after that stop. I sang "Working the Road to Glory" so loud it woke up the dogs, ponies and fighting chickens in the neighborhood. I was given a grand escort up a fairly smooth hill by the ever-present barking dogs. At the top of the hill, at an elevation of 1140 feet, I could find no tree marks. I set my compass into the earth, and it pointed toward High Point Mountain. It led me into Cave Creek Valley, where I found no sign of the survey.

I encountered a farmer with his Massey Ferguson stuck in a pond. He was surprised to see me emerge from the woods and volunteer to help. We walked to the nearby farmhouse where we secured a come-along.

We stopped in the kitchen and he poured me hot JFG coffee. "Boy, I've missed my JFG." It took me back to ranger-station staff meetings.

As we walked back to the pond, I told him about a time long ago when I too was stuck.

"Were you driving a tractor, mister?"

"No, sir. My life was stuck seeking after worldly things. I asked my precious Lord to save me from my mud hole. Now if I get stuck, my Savior is there for me."

"Mister, that's a wonderful story! Let's me and you go get that Massey unstuck with this here snatch block. Tell you what, mister, when you're back this way, stop for some more coffee."

A warm pleasant feeling washed over me as I left the farm. This friendly attitude had been my impression of all the folks of Roane County. At Pine Grove, I climbed up Cave Creek Road to High Point near Roane/Loudon County line. Near High Point the compass led me to a knoll at 1200-feet elevation. I could only find an old logged-over area. As I walked into Loudon County, the houses were getting denser. I walked through a field of sage grass, brush and briars. Passing an old barn, I started to continue, but something caught my attention out of the corner of my eye. Just inside the door sat an old airplane. It looked like Wilbur Wright had parked it long ago. Not wanting to trespass, I continued on my walk to the Smokies. Rangers are taught to show respect for private property. The two priorities for rangers are to protect people to safeguard human life and to protect public resources—whether wildlife or property.

I came to an area that had been logged over and was brushy and flat. I walked under high-tension lines that were home to some crows. A red-tailed hawk appeared overhead, and the murder of crows gave pursuit. My compass led me to the Hines Valley Store.

I bought several moon pies and big orange dopes. With a sugar high, I climbed a steep mountain called Black Oak. At the top I found an old black oak with an axemark.

The compass pointed me to Interstate 75 about 1 mile away. At the interstate highway I could hardly hear myself think. I had to travel north to an underpass to cross to the other side. To get back on track, I had noted that my compass had pointed me toward three big oaks on the other side of the highway.

Three large oaks on line—
12 miles from Fort Southwest Point

Arriving at the oaks, I found the three trees on line and in a neat row. I noted that this location was 12 miles from Fort Southwest Point. I stood in front of one oak and set my compass on S76°E. The compass was looking directly at the other two oaks. *Why am I finding obvious waypoints every six miles? Or am I imagining things?*

Church just west of three large oak trees

There was an old white church with a tin roof to the right of the three trees. It had one door with high steps leading up to it. Nearby was an old corncrib that was tilted precariously. I leaned back against the big oak and fell fast asleep. The wind was stirring in the old trees when I awoke. I left a modest offering with a note under the church door.

I am a wayfaring stranger traveling my path. Pray for my safe passage if it pleases you.

        —Vinn.

I struck out again, following my compass toward Lenoir City. After a slow climb up to Chestnut Ridge, I descended toward Silver Ridge. Although I'm sure the 1100-foot peak of Chestnut Ridge had to be a waypoint on the survey, it contained no new clues, because of the human occupation evidenced by numerous houses lining the crest.

***

Lenoir City was visible from the high peak, and the compass pointed directly toward Bussell Island. As evening arrived, I walked into Lenoir City. The number-one priority was lodging—number two was a restaurant. I found a nice place to stay whereupon I stored my gear. I discovered Aunt Bee's Country Buffet at an approximate half-mile walk toward Loudon. It was the best food I had eaten in a while—beans, cornbread with a big slice of onion, cube steak smothered in gravy, and hot apple pie.

I wonder what the surveyors would have eaten. Buffalo steak? Trout? Venison? Who's to say I'm any better off? I sat long after supper, drinking hot coffee and reading the paper. After a pleasant walk back to the boarding house, I listened to some gospel music on the radio. A warm hot bath was welcome just before bed. I drifted off to sleep and slept well until . . . the deafening roar of a southbound freight train seemed to be coming through my bed at about two o'clock in the morning.

At breakfast the next day, I met some fishermen. They agreed to haul me over across Watts Barr Lake to Bussell Island. It was just to the southwest of where Hawkins line crossed Yarberry Peninsula. I had been eager not to leave the area until I had prowled that island. I couldn't help myself.

The boat rocked back and forth in the strong current. As we crossed the Watts Barr Lake, I observed that the Tennessee River was brown and silty, but the Little Tennessee River, on the other hand, was cold, clear, blue and fast-flowing. This difference creates an intermingling of the two colors where the two rivers join.

They dropped me off where the island tapers to a comma with two streams of the Little Tennessee River to my right and

Bussell Island (Coste), slightly southwest of Hawkins Line
at junction of Tennessee and Little Tennessee Rivers
Photograph courtesy of the Tennessee Valley Authority

left. I offered to pay when we arrived on Bussell, but they refused. This had been my impression of all the folks of Loudon County— friendly, helpful and courteous.

From where I was standing at the island tip, I could look upstream at the Tennessee River. The Little Tennessee River's larger and deeper stream was on the right; the left stream was smaller. I walked the length of Bussell Island up the Little Tennessee River. The island had been converted to grazing and pastureland. My pacing and chaining off the distance told me the island was 75 chains, or 5000 feet long. I walked the maximum width of the island and found it was 30 chains wide, or 2000 feet.

I found an ancient Indian mound at the center of the island. This was a mound of earth that the farmers had been mowing around for years, not so much out of respect but because the mowing machines couldn't drive over it. As I walked closer, I recognized a characteristic appearance of the low knob similar to the Indian mound at Bryson City, North Carolina, called Kituhwa, or as spoken in Cherokee, ga du wa. I observed an ample amount of flints and pieces of pottery.

As I walked back through the vast pastures, I was shocked to find an ultra-light airplane in the lower field. Further investigation showed a smooth runway headed upriver.

Bussell Island was first written of by Desoto's chronicler on July 2, 1540. Desoto called it Coste after the Spanish word Caste. He stayed here July 2, 1540, to July 9, 1540, and was given a green buffalo hide by the Indians. Desoto observed the Indians had gold, copper and silver; they told him it came from Chisca, near the present-day Boone, North Carolina. Desoto came to Coste from Chiaha—now Zimmerman Island in Douglas Lake near Dandridge, Tennessee—where he and his entourage stayed from June 5 to June 28, 1540.

A later journey by Juan Pardo in October 16, 1567, recovered chainmail of one of Desoto's soldiers. His travels had brought him to a village he called Chalahume (called by the Cherokee, "Chilhowee"). This was about 12 miles upstream on the Little Tennessee River from where I was standing on Bussell Island. Pardo traveled downstream two leagues—five miles—to the village of Satapo, the present Cherokee village of Citico. There he observed an Indian wearing Spanish armour. The Indian told Pardo he was from Coste, where the villagers had suffered atrocities under Desoto. Fearing reprisals from approaching villagers of Chiaha, Desoto had fled toward Alabama. The Indian had joined in the pursuit, catching up with him in Muscle Shoals. During the battle that ensued, the Indian took chainmail from a dead Spanish soldier. The chronicler indicated that an additional Indian mound was located on the Lenoir city side.

I was able to get a ride off the island with another fisherman. He hauled me over to the Yarberry Peninsula, where I located the Hawkins line about 100 feet upstream from Bussell Island.[1] I found a marked tree. I expected to find evidence of the line at that location, not because of my compass course, but because of information I had read in the *Roane County Historical Society.* The book said there is a hill visible from Bussell Island, and in turn, Bussell Island is visible from the hill. I had spotted the hill from the Lenoir City side of the Tennessee River when the fishermen took me across.

There I visited the Jackson Cemetery and found no indications of old surveys. At Yarberry, no boats or fishermen were available, so I managed to rent a large inner tube and floated back across the lake once more. I arrived at Beal's Bend Peninsula where I located a benchmark elevation 839 feet that was exactly on line. In the 1950s, the US Geological Survey did aerial topographical assessments, allowing them to establish benchmarks at given distances and elevations. At these sites were placed concrete posts with a brass plate on the top. This particular benchmark,

---

[1] Bussell island is now underwater (1975) by the impoundment of Tellico Lake.

coincidentally, fell directly on the Hawkins Line. At that point, I was 18 miles from Fort Southwest Point. I followed the compass line and emerged on Beal's bend. The compass pointed across Fort Loudon Lake to a creek next to Leeper Bluff. *I wonder if it is named for "leepers" into the water or a family named Leeper.*

I stood on the shore for two hours, looking warily at my slowly deflating inner tube. Then an angel in the form of an elderly gentleman came putt-putt-putting around the bend. He possessed a most beautiful l2-foot aluminum boat with a 3.5 horsepower motor.

"I will surely drown if I have to use the inner tube," I explained, adding that I needed to cross to Leeper Bluff.

"Why you want to go on the other side, mister?"

"I'm lost and I believe civilization is over there."

"I'll give you a ride, mister, but you sit right in the middle."

I sat very still as the kind man putt-putted us across. I had never been so glad to be back on solid ground. As the angel sped away, I noticed his boat was riding much higher in the water than when our two collective weights had come close to sinking it.

My compass, to my surprise, pointed me directly to the Blount/Loudon County dividing line. But before taking up my staff, I stashed my gear and walked south three-quarters of a mile to Cloyds Creek. That would have been Indian Territory in 1797. The surveyors would have found a grist mill and a sawmill. Now, I watched as speed boats came and went from Unitia boat ramp.

I returned to my gear and struck out on the S76°E course. The line followed directly on top of the county survey for 4900 feet— two surveys running together from different timeframes in history. Then the county line struck off toward Unitia. The stroll across pastures and open fields was pleasant. Gone were the mountains of Loudon County.

The area was becoming more densely populated. Lunch was at a small country store in Friendsville—a big slice of bologna and thick piece of cheese. This area produces some of the finest marble in America. Many folks offered me a ride as I walked. But I told them I must walk as many steps of the journey as I could. I visited the site of the First Friends Church, built in 1794. The overgrown remnants of the old building were evident. The cool spring the Friends had used was still there but did not seem to be used now.

The evening grew close but I chose to continue my journey to seek an isolated spot to camp. My compass pointed me toward the northeast end of a mountain called Grey Ridge. The sun was setting as I paused on a pleasant knoll for the night. I seemed to be in a series of knolls and ridges.

Early the next morning, I studied my map and then prowled around the area and located Lane Spring. It was a pretty good-sized spring trickling down from a hill. I spotted a large rock with a carving on it. I was 24 miles from Fort Southwest Point. With a stick of chalk, I highlighted a crowfoot carving that was on the rock.

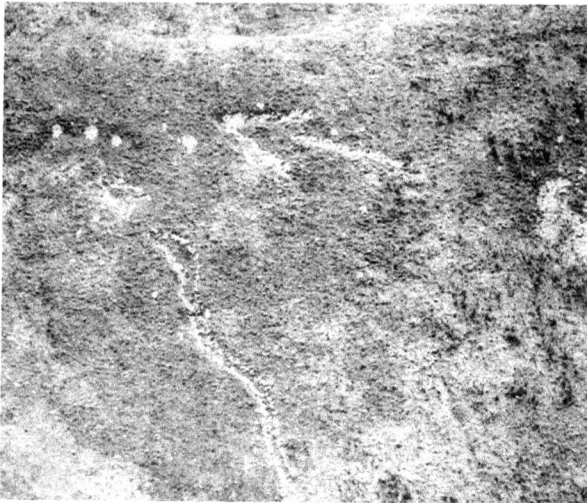

Rock carving near Lane Spring,
Friendsville, Tennessee

The crowfoot carving always points toward where a witness tree or stone bound can be found. I found four drill holes in front of the crowfoot which meant the Hawkins Line was four chains in the direction the crowfoot points. I followed that direction and paced off four chains and found myself standing in my camp. That is where, on the day before, my Jacobs staff compass said the Hawkins Line would be. The area had been cleared for farming so there were no longer trees to represent the witness trees or piles of stones to represent stone bounds. These terms were lingo of Washington's and Jefferson's surveyors. The adjective "witness" referred to signs that came in series of threes—three axemarks, three drill holes in rocks and other symbols. The trees were called witness trees because they served to hide the survey in front of everybody. The stones were stacked in a circular formation in an interlaced manner.

I packed my gear and set the compass in the soil. Sighting S76°E, I was looking at a mountain two-and-one-half miles away. It looked like a snail minus its shell. I traveled for about three hours across pasture and plowed fields. The compass then pointed me toward a paved road. I arrived at the base of "Snail" Mountain and was standing at Big Springs Road. *Man. I could use a cool drink from those big springs.* I looked ahead and spotted a white house to the right of a group of trees. I set my compass once again in the earth. Sighting down it, I was looking down a line of three large oak trees.

These were as large as the ones at the 12-mile location. *This cannot be a common occurrence.* I believed those places with three trees close together—all right on the survey—indicated a crude "witness tree." Instead of a tree with three axemarks on it, I kept finding, often at six-mile intervals, three trees in row-formation on line! These would seem unimportant to someone seeking to destroy the survey—property owners who might find themselves on the "wrong side" of the line. I believe the 1797 surveyors hid the line in plain sight. I had developed a keener respect for Benjamin Hawkins. There I stood—looking at a coded message from the past.

Three big trees on line, Big Springs Road,
Maryville, Tennessee

The resident of the white house emerged on the front porch. His wife viewed me with caution from the safety of the screen door. My big smile and wave of the hand dispelled their concern.

"How long you had these big trees, sir?"

"God owns those trees, mister. He planted them way before any white man seen them. You like trees, mister, or are you been in the sun too long?"

"Well, sir, I just ponder why they are in a row and spaced the same distance apart."

"Well, mister, with all the trouble we have in the world today, your pondering over a tree surely won't cause much more trouble! Me, myself and I got cows to tend to."

With that I was left staring at the trees, which I continued to do for ten minutes, and then I followed my compass over a small hill. For twenty minutes I walked through cows and nothing else. At the top I set my compass, and it pointed toward a mountain that looked like a golf course. Hours later I was on the 12th tee looking rather silly.

"Where's your clubs?"

"Just passing through, sir!" There was nothing but green grass on these hills. So I planted my compass and looked down it.

There, far away, was Chilhowee Mountain. It was on the Chilhowee Range, which runs twenty-seven miles from Sevierville to the Little Tennessee River. This mountain looked very different from the others nearby. For one thing, it had a cone shape, and it stood out by itself. The distance appeared to be 12 miles. I got out my map and could see clearly the survey came over a place called Blockhouse and continued on to the high mountain—elevation 2062 feet. I set out on my course and soon crossed Highway 411.

On the other side of 411, my route was blocked by Maroney's Restaurant. So the only reasonable thing to do was to eat my way through the obstacle. I consumed a large cheeseburger, fries and a large chocolate shake. The shake was made the old way with real ice cream. I still don't know how the blades on that stirring machine work. The hamburger was made the old way with just a hint of pepper on it. They graciously permitted me to exit through the back door.

I set out for Chilhowee Mountain. Shortly, I arrived at Pearson Spring and cooled myself. At this point, I was 30 miles from Fort Southwest Point. *I bet this would have been where the survey party ate lunch.* I crossed Carpenters Grade Road and went through more neighborhoods. Soon I came to Montvale Road—more neighborhoods.

I arrived at Pea Ridge Road, and as I stood there, the Blount County Deputy Sheriff drove up. He said some folks were concerned over my presence. No one walks anymore; it's just not in style. I showed him my journal and also showed him I was a brother of the badge. He said with a slight grin, "Oh, I see you're a possum cop."

Oh, that nickname is nothing, I thought. *CBers call us Pine Pigs.* He contacted someone who verified my credentials.

I crossed Peach Orchard Road and came to White Oak Springs. The water was cold and clear. I finally was near Blockhouse. It was named for Joseph Black's blockhouse, which was built March 13, 1794. A blockhouse was built on the boundary. An octagonal structure made out of poles, it afforded the farmers a place to flee to security in an Indian attack. There were little slots that could be used to shoot through. The eight sides allowed a view of other sides from any window. The surveyors must surely have stopped here, for if my compass was right, the blockhouse would have been in Indian Territory—the white people, later to be relocated north of the line.

***

I finally left the confines of a fairly populated area and entered a more mountainous environment. *I'm Home.* On Coon Hollow Knob, I found the telltale signs I was used to—blazes and axemarks. The next ridge on line was Little Mountain—elevation 1220 feet. I descended into Coon Hollow and a large field at the junction of South Fork and Mook Creeks. I crossed Butterfly Gap Road and climbed over a small mountain—elevation 1140 feet. Down the other side I came to beautiful Butterfly Gap Church. I rested on the front porch and felt welcome there. It was time for a song.

> *I am a poor-Wayfaring stranger*
> *Wandering through-This world of woe,*
> *But there's no sorrow, Toil, Nor danger*
> *In that bright land- To which I go*

I'm going there- To see my family
I'm going there- No more to roam
I'm just a -going over Jordan
I'm only going - Over home

Surely the presence of the Lord rests here. I felt safe from my journey. I would love to have left something, but the hamburger took my last coin. I would need to replenish my funds later. I wondered if these various buildings along the way were coincidences, or were they intentionally built at or near the old survey line . . . *or am I just imagining things?*

I had a rough climb up to the top of Little Mountain. The next feature of the route was an obstacle course—rocky terrain, boulders strewn, briars overgrown, fallen timber. It hurt the body in every place to crawl over, around and through it. After a brief stop at the top, with gear in tow, I descended into Greasy Cove. At the bottom, my compass pointed directly at Hise Roulette Spring. If no one were to disturb me, I planned to camp near the spring.

The water was crystal clear and good. The wind howled over the Chilhowees as I drifted off to sleep. This had been an eventful day. It seemed I drank from every spring I came to. That was a good thing as these springs issued forth from the side of the mountain. The water is pure and sweet—it can't be polluted. Streams do not flow into them but the springs flow into streams.

The climb next morning was a rough one—more briars and boulders. About ten a.m. I crossed the Foothills Parkway, which had just been completed a few years earlier. The compass pointed up the steep bank on the other side of the road. Thirty minutes later I topped out on the cone-shaped mountain—elevation 2062 feet. I was 36 miles from Fort Southwest Point. This mountain was high and dry, and no trails or roads went through it.

I found a wind-protected area a short way down from the top. It was a peaceful place. I decided to camp for several days. I set up my camp with the protection of some huge boulders. If a tree were to fall, the boulders would keep it off me. I lounged around most of the day eating—huckleberries and other wild forage—and sleeping. The warmth of late spring was all about, with laurel and azaleas blooming. This mountain was south-facing with pine and oak trees. In the evening, I went back up to the top and found some big oak trees with axemarks forming a circle around a large, flat shelf rock.

For thirty minutes, I removed fallen timber from the rock. Then I spent another thirty minutes sweeping leaves and muck from it. There on the center of the rock was my reward—a clear triangle with one point heading toward Hesse Creek. I set up my compass on top of the triangle. Sure enough, the compass pointed toward a big oak with a blaze. I was on the line. From the triangle rock, they would have shot back to Hise Roulette Spring.

Triangle Rock on line
Elevation 2062 feet

The line tree was about 60 feet away. I moved to the other side of the tree and found another blaze on that side. Setting up my compass, I could see clearly that the old survey line crossed below an old farmhouse at Blackberry Farm in the valley below, based on The History of Blount County, written by Inez Burns.

The Hawkins survey in 1797 indicated there were four "intruders" in this valley. These were white individuals living inside Indian territory. One was the builder of a log cabin where the old farmhouse now stands at Blackberry Farm. That builder would have been John Hesse who was a Swiss immigrant and for whom Hesse Creek was named. The cabin was 300 yards inside Indian territory. The reason was that the only cool, clear spring was located there. Hesse chose to build inside Indian territory rather than where he should. Today, the spa at Blackberry farm sits atop the remnants of the Hesse cabin. In October 1797, six months after the survey went through this area, Hawkins returned with troops to forcibly eject the intruders.

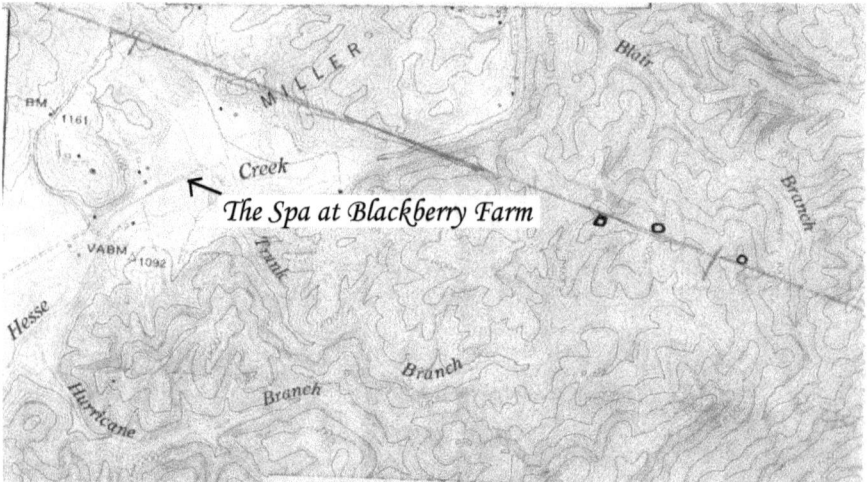

Map showing Hawkins Line crossing Blackberry Farm at the spa

I busied myself chalking and photographing the bearing trees around the triangle rock. A bearing tree is more recognizable than a line tree in that it contains the actual line course embedded in the axemarks. Next, I stored my gear and went in pursuit of a spring. I could hear the gurgle of water down the ridge a ways on the right. In a rhododendron hollow, I found a spring issuing out of the rock. It took awhile to fill my vessels.

As I climbed back up to my camp, I noticed a bright sparkling rock. It looked like gold! *I will come again to this area.*

Storing my water, I revisited my line at the top of the mountain—elevation 2062 feet. The compass pointed S76°E from the triangle rock to a large oak tree three poles away. One pole equals 16 ½ feet, so the first tree was 49 ½ feet from the triangle. The first tree had to be a witness tree. It had three blazes cut to form a triangle or pyramid. The top of the triangle pointed toward two other large oaks on line. I paced off another three poles to the first sight or line tree.

Blaze

Hesse Creek at Blackberry Farm

Clockwise: 1. Three-blaze-triangle witness tree 2. Three-notch line tree 3. Carved line tree with owl

The first line tree had three notches on it. I paced off another three poles to the second sight tree. This tree looked directly toward Hesse Creek. The tree had numerous carvings on it. One was a long vertical axe blaze. This was consistent with other sight trees I'd found. They all had only one blaze, either horizontal or vertical. Next to the vertical blaze, I could see the form of a great horned owl. Above the owl were numerous notches. At the very base below all these was a crowfoot carving pointing toward Hesse Creek. These old crowfoot carvings resembled a pointer or arrow symbol. Just past the crowfoot on the line at a little knoll, I found an area where the Indians accompanying Hawkins may have paused, camped and made bird points and arrowheads. The ones I found were the Kirk corner notched points, a typical arrowhead widely recovered in North Carolina. While searching in the area for springs, I found the quarry, the probable source of the cream-colored flint used in the Kirk-point arrowheads. It was one-quarter mile west of the arrowhead "factory" on the triangle-rock mountain.

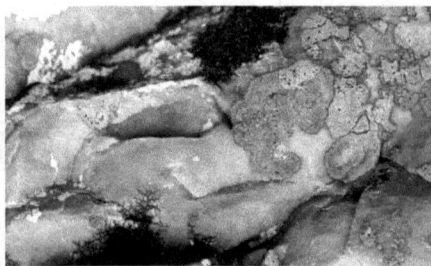

Photograph on left:
Flint quarry at Triangle Mountain

Photograph on right:
Cream-colored flint for making
arrowheads

Back on the line, I found a rock medicine bowl, indicating an older presence of the Indians sometime in the past—as long ago as 8000 years. It had been chiseled in the solid rock by deer-antler awls because I could see grooves characteristic of those made by the awl pecking into the rock. A club made out of chittum wood typically would have been used to hammer the awls into the rock. Perhaps the Cherokee with Hawkins were already aware of that location or recognized the medicine bowl—coincidentally, right on Hawkins Line! The Cherokee would likely have been nervous, knowing the survey was headed so close to their arrowhead factory.

It appeared from what I'd found so far that the Hawkins Survey used the Meets & Bounds method to survey the line. Meets means to report the line according to items encountered—like an unusual rock outcrop, mountaintop or tree. Bounds means to report the line according to natural topographical phenomena that were followed—like a stream or ridge. Hawkins also blazed witness trees, drilled triangles in rock outcrops, built stone bounds, and planted three oak trees in a line. All these markers were between one mile to six miles apart. I had found sight trees a minimum of every half-mile between the witness trees. A bearing tree was usually located at the top of a mountain. A bearing tree was always surrounded by three witness trees. A witness tree had one blaze or notch facing the bearing tree.

The surveyor was referred to as the level man or compass man. He would stand at the last point of sight established and sight to a point ahead, such as a tree on the line—hence the name bearing tree or sight tree. Then the rod man or pole man went to stand at that point. The pole used was 16 ½ feet long or the length of one rod, pole or perch. A chain, 66 feet, was four poles. The axemen blazed the line by notching axemarks on each tree on the line between where the level man and the rod man were standing. When the axemen arrived at the rod man, they blazed witness trees around the bearing tree where the rod man stood. Then a reverse azimuth was shot back, in this case N76°W, to the former sight or bearing tree to confirm the line. A surveyor's compass is divided into four quadrants. In the quadrants are SW, NW, SE and NE. The lines dividing the quadrants, labeled 0 to 90°, are at right angles to each other. This allows the back-sight number to equal the forward-sight number.

It appeared my next high point was a ridge about two miles away—elevation 1560 feet. This high point would divide Trunk Branch and Blair Branch. I ceased my project for the evening and returned to my camp. After a fairly good meal, I got my warm coat on and sat in a rock formation that resembled an easy chair. Hot cowboy coffee, a warm coat and a good easy chair was my reward this day. The view of Gregory Bald was spectacular as the setting sun transformed the side of Hannah Mountain into gold. I woke up later still sitting in my easy chair. The night was pitch dark as I crawled into my tent.

"Easy chair"
Rock, Triangle Mountain
Elevation 2062 feet at Blackberry Farm

The ospreys were a concert for my ears the next morning. I fixed more cowboy coffee and donned my coat to sit in the easy chair once more. The sun peeked over Gregory with a brilliance more sublime than I could remember. I packed my gear and returned to elevation 2062. *I wonder what this mountain's name is? I believe I will call it Triangle Mountain.*

I set my compass at the #2 sight tree and followed it off the side of the mountain. The slope was very steep as I descended. Halfway down I again came to the pretty colored rock—this time it was bright red. I recalled that just west of there was Gold Mine Creek that flowed into Hesse Creek. The history books document a trip into this area on October 14, 1567, by Juan Pardo. It was the second time a white man ever explored deep into the American wilderness. His expert on minerals located a reddish ore that had gold in it. I remember he wrote the following:

> On top of the high mountain they reached at the end of the day, Juan Pardo found a small reddish stone which he gave to Andres Suarez, who was a "melter of gold and silver."

The gold mine is currently under the man-made lake called Top of the World. A large amount of iron ore at the site supplied raw ore for the Amerine Furnace and the Carson Iron Works. The remnants of the Amerine Furnace, a forge built in 1840, was located on Hesse Creek at Mountain Homes. The Carson Iron Works ruins are located one mile above the Abram Creek campground in Happy Valley, Tennessee. The iron ore was 50% pure and was called limonite. Gold is present in quartz veins that occur in the 40-foot-deep limonite ore.

Coming into Millers Cove, I crossed two fields and an old home site. The home site contained some huge walnut trees that surely dated back to 1797. Having crossed Hesse Creek, I climbed Walker Mountain.

I had walked two miles, and my instincts told me to look about. Slightly to the right of where I stood was a huge oak.

The oak was 46 inches in diameter at breast height (DBH) and contained a crowfoot and a huge axemark that had rotted out a hole big enough for a bear to live inside. The crowfoot pointed S76°E. The tree also had one notch indicating it was a line tree. There were indications that a large cat regularly sharpened his claws on the tree.

Both photographs:
Forty-six inch oak with huge,
rotted-out axemark and crowfoot,
Walker Mountain Blackberry Farm

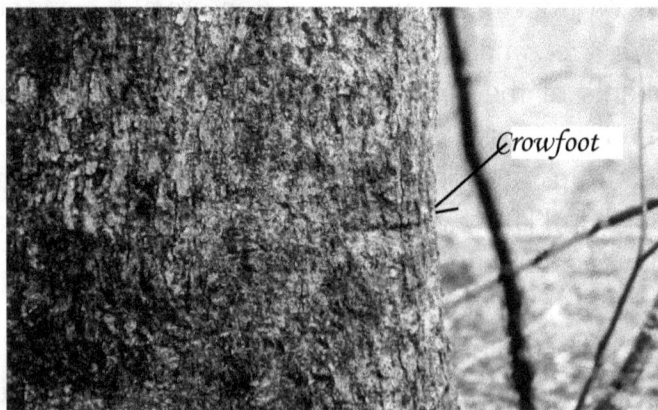

Crowfoot

I continued following the line for 12 chains, and there in the middle of my path was a 42-inch (DBH) oak with one notch. There were other marks on the tree that could not be interpreted.

An old tree, as it grows over the years, might elongate a letter, number or symbol so that it could no longer be recognized. The elevation was 1480 feet. I followed my compass for another 24 chains, or 1584 feet, to elevation 1560 feet.

I was at a point where Walker Mountain corners and heads toward Hurricane Mountain and the Great Smoky Mountain Park boundary. Directly in front of me flowed the narrow gorge of Blair Branch. I managed to locate a dirt bound right on top of the knoll. A rock/dirt bound is a man-made formation—a surveyor's tool— composed of layers of stacked rocks and piled dirt. A dirt bound or an earth bound is mostly dirt, and a rock bound is mostly rocks. There were only young dead pines there—none old enough to have been alive in Hawkins' time. But to my left, approximately 60 feet away, stood a huge half-dead oak.

Huge half-dead oak
with notches,
Walker Mountain
Blackberry Farm

The three trees are
depicted as on-line
circles on
map, page 28

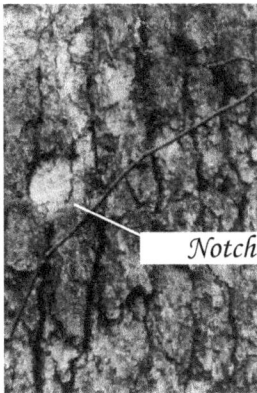

Notches

I quickly traveled to it and found, about 20 feet up on the tree, the trunk had three notches cut into two sides and the remnants of other notches on the third side. The axeman would have had to climb up the tree to mark it because, even centuries later, whatever is marked on the face of the tree remains at that level because the tree grows from the inside out. The top of the old tree lay downhill and the remaining trunk was barely alive. I believed this might actually be the witness tree pointing to the earth bound up on the knoll. One seldom finds earth bounds serving as a bearing tree. Likewise, one seldom finds witness trees that lack any bearing tree nearby. The reason this witness tree was standing solo, I surmised, was that this area was logged over and over. Whereas the witness trees may be legal to cut, it is illegal to cut or deface a bearing tree. This tree had a DBH of 47 inches and was nearly on line. The tree appeared to have waited for me to come along before it gave up the ghost and went to tree heaven. I was about two miles from my Triangle Mountain and was not looking forward to staying any longer than necessary in the dark deep gorge of Blair. I elected to cut straight across Blair Branch and shoot for a deep gap that I could see ahead. Large beech trees dotted the steep hillsides of Blair Branch. Many had old carvings left long ago but, alas, no Hawkins or Pickens traces did I find. I couldn't escape the feeling of being watched and was so vexed, I left without fully investigating the area. *I will bring twenty members of my church with me next time. They can watch my back and fight off any haunts and ghosts.*

Climbing out of Blair Branch, I came to a pleasant gap— elevation 1380 feet. Looking back I saw the steep, foreboding gorge of Blair. Ahead I saw smooth walking down the right fork of Davis Branch. This nearly flat valley had to be an area where the Cherokee would have camped. I had a pleasant walk as I sang "Waltzing Matilda." Far up on the hillside a woodpecker laughed at my singing. *You would think there would be no critics in the woods*. At a point halfway down the pleasant hollow, the survey line rose to a ridge. Climbing up to the top, I came to a narrow hogback or spine. There was nothing there to indicate the line, so I dropped down the steep slope to Davis Branch. There I found the remnants of the Little River Railroad bed.

The rail line went to Ace Gap on my right. From Ace Gap it dropped into Hesse Creek, where once it had traversed several valleys logged by Little River Lumber Company. There at Shell Branch were the remnants of the Eldorado Logging Camp. The area was probably logged sometime around 1911. Now the area was sort of a no-man's land with very few trails to the Eldorado area. I looked for signs of the survey line but found none. Most likely, the logging operation removed any indications.

The compass led me up a little hollow eastward toward Verified Altitudinal Bench Mark (VABM) 1605. Hawkins line crossed just south of the VABM. I first encountered a huge white oak tree with two big axemarks facing toward Dunn Spring.

Huge white oak with two axemarks and rotted-out hole facing Dunn Springs
Location VABM 1605

The bottom axemark had rotted out a hole in the tree, which now was not only big enough for a bear . . . it contained one! He was awake and staring at me. Now I knew the source of my eerie feeling. When I had been a back-country ranger, I met bears on a daily basis. I realized this bear was not a threat to me. I know not to turn and run because to do so would trigger the predator-prey instinct in any animal, be it dog, bear, mouse, grouse or human. A fleeing subject is prey. Instead, I faced the bear and backed away. In some countries known for attacks by lions and tigers, natives walk through the wilderness areas with masks on the back of their heads to give the impression that they are facing the animal as they depart. With great dexterity I successfully balanced the confrontation with the bear and my determination to photograph the tree. Thus I escaped between the horns of a dilemma.

After a safe distance, I stood with the bear tree at my back, and sighted down the compass. I was looking toward another oak. I followed the line three chains to that tree, which had a vertical axemark on the upper side. Standing in front of the axemark, my compass pointed toward Dunn Spring in Dry Valley.

After photographing the tree, I set the compass in the ground and followed where it pointed. Down a steep slope, I descended into Kinzel Branch, which flows not into Hesse Creek but Little River. At the branch, there were no signs of the survey, so I continued on my journey. Up, up a mountain I climbed and at the top came out on a good road. All about me were summer cabins, A-frames and other structures. Ahead of me I saw Laurel Valley Golf Course and its nearby lake. Dropping down into Laurel Valley, the line passed more cabins. I descended into an area below a dam on Tipton Branch. There were few trees in this area that would have axemarks on them because of the clearing of the land for farming. After a time, I crossed Short Creek and came to the Dry Valley Road.

The Tuckaleechee Methodist Church was on my left, and Dunn Spring Trout Pond was on my right. At Dunn Spring I was 42 miles from Fort Southwest Point en route to the Great Smoky Mountains. I could not help think these six-mile intervals had to have something to do with dividing up the land—The Revolutionary War land grants.

I garnered some concerned stares from Boat Gunnel Road residents as I proceeded toward the north side of Little Mountain. The large trees were few and far between as I climbed the slopes of Rocky Knob.

At one point, the Townsend Elementary School was visible on my left and children were at play in the foreground. I believe Hawkins would have been here around mid-May, 1797. Hawkins' report stated that two intruders were living on the Indian side of the line. These two intruders would have lived on my right in Dry Valley or Cades Cove. Down I descended into the Hollow of the Fox then up a big fat mountain that overlooked Mitchell Hollow. Down again I went to the hollow named after John Mitchell, who settled here from Ireland in the mid-1800s.

The day had been long as I climbed up to Fall Ridge. At the west side of the ridge, I was pleasantly surprised to meet the familiar boundary signs of the Great Smoky Mountains National Park at Chestnut Mountain. *I am home! This means no more barking dogs. No more critical stares from local residents. No more being chased by bulls, cows, billy goats and other sundry animals.* I was caught by the fact that there before me was a corner tree. The park survey on my right went toward Bryant Gap and Schoolhouse Gap, and the other line, going toward the Tremont area, was directly on my compass reading S76ºE. *This has to be part of the original survey line in front of me. At this point, the park boundary survey of 1920 and the historic survey of 1797 are superimposed.*

I will rest here for the night. The little hill is peaceful.

***

That evening about 6:30 p.m., the bells of Bethel Church started ringing. I assumed that it must be Wednesday evening service. I could see the cars arriving in the lot and people threading into church. Oh, if I were not so tired, I would gladly walk the distance. After supper I read far into the night from my pocket Bible. The light from my candles provided ample illumination. It occurred to me that, as I prepared for bed at ten p.m., the folks down there must surely see the small flickering lights. *I wonder if they will think it's the Chestnut Mountain lights being carried by the ancestors of those living here.*

I slept late into the morning. I took care of laundry chores and hung my sleeping bag in the sun to air out. With soap in hand, I visited Fall Branch and located a small waterfall to take an invigorating shower. These mountain springs never warmed up, it seemed. Returning to camp I lay in the hot sun all day. I woke up as the evening sun was dipping in the sky and decided to spend another night. I prepared fried corn cakes and a big pot of stew from my remaining supplies. I ate as I have never eaten before—like a ravenous wolf.

The next morning I was up with the sunshine, eating flapjacks and drinking hot cowboy coffee. I made a trailside coffeemaker out of a #10 pork-and-beans can by punching two holes near the top and inserting a wire to suspend the container over the fire on a triangle frame of sticks. After bringing the water to a rolling boil, I poured in coarse ground coffee and allowed it to cook for just the right amount of time. Then I added cold water to settle the grounds to the bottom and dipped little tin cups into the rich brown liquid.

As I packed my equipment, I noted it was much lighter from the previous day. I planted my compass in Mother Terrafirma and it pointed right down the park service boundary line. My map told me I was headed toward the Chestnut Top Trail and the Tremont

Wye. The first marked tree I came to was about 200 yards up the line. It was a huge poplar tree where the axe blaze of 1797 had rotted out a large opening.

Rotted out axe-blazed tree. Location Chestnut Top

The opening was roomy enough for a bear to live in . . . and sure enough, one was living there. The surprised bear came out and bluff-charged me. I shouted and blew whistles and whacked the bushes. With much bluster from me and the bear, it was decided I could photograph the tree from afar. But the bear, at the last moment, decided my little blue pack had to be inspected for any illegal items. Trouble was the bear was an amateur with zippers. He made me a new opening in my pack. This was very thoughtful of the bear, for I had always wanted better access to my pack.

I quickly planted my compass, and it led me away from the bear. Farther down the line I came to what I called the Bible tree. Within a large blaze on the tree was carved a small Bible. I thanked my precious Lord for the Bible symbol, for I had fallen shamefully behind in my daily worship. Farther down the line I

Bible-blazed line tree, Chestnut Top

came to another line tree. A groove had been cut into the tree, resembling a big smile. I passed several other trees with notches and axemarks similar to many I had seen along the way. Just when I might have started to get bored, I came upon a line tree with a baseball field shape. Hawkins was surely innovative in his methods of marking his survey.

Left: Smile on line tree, Chestnut Top
Right: Baseball field shape line tree, Chestnut Top

As I topped out on a ridge, I could see the Chestnut Trail below me to the right. It was obvious I would soon merge with it, but just before the merge, I noticed a stack of stones. These stones had been in place for a very long time. As I gently uncovered decades of debris, the letter L appeared on top of one of the rocks. *Does*

Stone bound with "L" carved on rock, Chestnut Top

it indicate "Line" or is it an Indian symbol? I photographed it and walked out onto Chestnut Top Trail. The trail curved sharply around the ridge, and the boundary line for the park took off in a different direction. But, for about a mile, we had merged a modern survey with a survey laid down when our country was brand new.

I started down the ridge toward Tremont Wye. A wye is a direction change in a railroad or waterway where the option exists to go either left or right. In the railroad systems, the wye is configured so the train can go either way, both on leaving and on returning. There is a short track crossing before the intersection, facilitating the shift in direction. The incline got steeper and steeper. I could hear the roar of the river as I held onto trees to get down. I came crashing through the laurel and rhododendron at the big hole of water at Tremont Bridge. There used to be railroad tracks there. They had long since been replaced by motor roadways. Visitors grabbed their children, and picnickers grabbed their food, fearing this wild apparition that had just emerged from the woods as if from nowhere.

I assured them I was a park ranger on vacation, but they believed me not. In their haste to leave, they left a big bag of corn chips near a rock. These, I swear, were the best corn chips I had ever eaten. I was 45 miles into a 60-mile hike. It was time I rested up and resupplied my food. I marked the spot where I would resume my journey later. I walked toward Townsend, about one mile downstream. At the Townsend Wye, tubers and swimmers were active in the sunshine.

I paused at the site of the old check-in station for fishermen. It was gone now, but a long time ago it was a nice place to work. I learned several lessons from those wonderful patriots of the mountains. For those folks who wished to camp in the backcountry, the check-in station was where we got a campfire permit. I got to meet all the anglers who fished the day I went in.

I fondly recalled one fisherman from the area who always caught the largest trout. I watched him several times to see if he was using any illegal items. He always used flies and lures that were permissible. One day, when I thought I was sneaking up on him, I suddenly had the feeling something was behind me. I turned to see a pair of eyes looking out of a bush. "You been on my back trail, Ranger!" he said.

"Yes, sir, I have, but after watching you for some time, I'm giving you a clean bill of health. By the way, sir, how did you know I was behind you?"

"Well, Ranger, a man that stands six foot seven and weights 260 causes quite a commotion. Birds fly away in panic, tree limbs break and river rocks clack together when you walk."

I stopped watching him from that day on.

I walked down the river toward Townsend, singing an old mountaineer's song. The song told of early settlers' working to maintain the wagon roads of the Smokies.

> *Working the road to glory*
> *Working the road to glory*
> *Helping the weak and the blind*
> *I want to hear them say, Up there in glory land way*
> *You made it easy, for those behind*
>
> *I often think of the days, So swiftly passing away*
> *The road to heaven is hard to find*
> *So I am doing my best, To reach that home of the blest*
> *And make it easy for those behind*
>
> *Working the road to glory*
> *Working the road to glory*
> *The road to heaven is straight, and so narrow the gate*
> *Along there's trouble of every kind*
> *I want to hear them say, Up there in glory land say*
> *You made it easy for those behind.*

The Little River Campground came into view. There before me, an audience of campers had gathered to hear the song I sang. I rented a campsite on the river near a big swimming hole. For the next three days, I relaxed, did my laundry and purchased food. I also enjoyed meeting and greeting folks from all over the nation who were staying there. Every evening would be spent around a campfire swapping stories. During the heat of the day, I would swim in the pool or the river. The river had some sizable small-mouth bass. The big swimming hole was full of rainbow trout also.

Sweet Julia joined me the third day at the appointed time. She brought me many sweets made with her loving hands. She had been occupied in my absence with her duties as a first-grade teacher. We talked of school and the summer break, which was rapidly approaching in a couple of weeks. I told her about dogs, bears, people and an airplane, and shared my discovery that many of the old survey points still exist. The days apart had been long and lonely. We were so glad to be reunited. The special moments by the fire in our secluded campsite would provide an afterglow for many days.

The next morning she hauled me up to the Tremont Wye where I had paused on my journey. I passed along to Julia my journal and the intended route from there forward. I assured her I would be waiting at Collins Gap for her on June 6 when she was to pick me up. We bid farewell. I returned to my journey, and she returned to teaching.

I once again set my compass into Mother Terrafirma and followed it up a laurel slope. The line headed for Ashley Branch and crossed it at a concrete water tank. Most ranger stations have such a water reservoir that works on a gravity-flow system, because of the unavailability of wells and city water. To minimally disturb nature, water enters from a spring or stream, flows in the system through a sand-trap filter and exits at the bottom. Barring squirrels, salamanders or oak leaves, it is relatively pure. This water system supplied water for the Tremont ranger's house. I drank from the overflow which was cold and clear. I sidehilled, came across Bear Sign Branch, waded the river above the Cubby Hole—a lovers' tryst site—and walked into a whiskey still on the ridge above the river. There were milk jugs everywhere, a propane tank and burner.

Since I'm a civilian now, I believe I should sample the purity of the liquid. I justified my taste test as purely scientific in purpose. It burned all the way down and gave me a warm feeling. *Well, that took care of that persistent cough I have had lately.*

I continued my route walking my way to glory. I topped out on a lonely ridge with no signs of blazes or notches. Just when I was about to second-guess my compass, there, right in front of me, was a large oak. A crowfoot with three notches was visible on the tree. The crowfoot pointed up a ridge.

Line tree with crowfoot and three notches, Lumber Ridge

Blazed line tree, Lumber Ridge

I followed it across Lumber Ridge Trail, and, after climbing three chains, arrived at a blaze tree. The blaze tree pointed uphill to a large rock outcrop. At first, I saw nothing at the rock outcrop. But as I started to walk away, there, out of the comer of my eye, I saw the letter P. The P was chiseled into solid rock. The P had to stand for Pickens—General Andrew Pickens. Thusly, I named the site "Pickens Rock." I counted my total chains walked and found the P was 48 miles from Fort Southwest Point. The bearing was S76°E right over the P. Again, I was impressed with these statesmen, heroes, and all-around respectable men—Hawkins, Pickens and Tim Barnard.

On-line letter "P" carved on rock, Lumber Ridge

I will stay here for the night, for the burden I bear is heavy with cornmeal, bacon and canned goods. I climbed to a point above Pickens Rock on the main Mill Ridge—elevation 2480 feet. Looking back along the survey, I could clearly see the Foothills Parkway at elevation 2062. *What a view! And it will only get better as I gain altitude.*

Pickens Rock view

I pitched my camp on the flat ridge above Pickens Rock. The ridge was covered mostly by pine trees in varying degrees of dying from beetle infestation. The wind-swept ridge was a pleasant place. I could, at times, hear birds and animals on the Pigpen Branch side as well as the Spruce Flats Branch side. I went back down to Pickens Rock and, for a long time, viewed the sun setting over the Chilhowees.

It occurred to me that Pickens Rock made a very nice couch of sorts, and that was the last thought I had. I was awakened about 10:30 p.m. Something was rattling my pots and pans at my camp. I hurriedly responded to the ridge to find the butt of a bear sticking out of my tent. Momentarily lapsing into a nonranger persona, I dealt with the bear stealing my food in a manner I was not proud of. With an oak limb, I proceeded to wear his butt out while I chased him through the woods. *Why is it that the bear always tears into my meal and flour sacks?*

Now I was fully certain that I was truly back in the Great Smokies Park; rangers deal with bear problems on a daily basis, especially in May and June. The bear chose not to return the rest of the night. When I woke up the next morning, I looked like I had rolled in flour. I took consolation from the fact that he had flour all over him, as well. *What to heck! I'll just wear it off over the next few days.* I ate breakfast and fixed a fresh pot of coffee to view the sunrise coming over Meigs Mountain. It was named after Return Jonathan Meigs, who resurveyed the Hawkins Line in 1802.

Up ahead I would encounter the point where his survey established a more accurate Cherokee Nation and U.S. boundary.

I stayed at my camp far into the morning—just relaxing. At 10:00 a.m. I started down the line toward Spruce Flats Branch. Crossing the creek, I found no blazes or other signs. I started climbing again as the compass pointed me toward Upper

Buckhorn Gap. As I passed the gap, I noticed the line crossed to the right of an old home site there.

Gone Home carving, Upper Buckhorn

A short distance beyond, I came to an old beech tree that had to be at least 300 years old. The tree held a message about a person who had passed on to the other side. I did my best to interpret the old inscription. Scientists would call it a pictograph. The person's name started with M, a beautifully carved letter. I couldn't make out the rest of the name. He died October 26, 1908, and he was born in 1839. A large B was used to clarify that the year given was the birth year. The person would have been 69 years old. The fact that he believed in God is of no doubt. The message indicated that the person went home to be with Jesus—there was a carving of a hand pointing to heaven. I felt the person could have died while working for Little River Lumber Company. The lumber company was active logging the area and building the railroad up the gorge in 1908. Besides the plain slate headstone, there was a marker behind the tree and a depression in the ground, indicating a grave. How interesting that the inscription was on the tree and not the stone!

***

I began my climb to the top of Meigs Mountain. Starting at elevation 2800 feet, the climb would be an altitude gain of 1160 feet to an elevation of 3960 feet. It took me most of the day, lumbering to the crest of the ridge—more rocks, briars and fallen timber . . . I was completely worn out when I got there. The ridge is big, precipitous, fat and wide. I drank plenty of water and ate some snacks. I didn't even pitch my tent but just curled up against my pack. I dreamed dreams of being a young ranger who could, if need be, run up these mountains and then dance all night. But alas, those days were gone, never to return. I spent two days recouping on top of the mountain.

I did manage to find a large oak with an axemark on it at the bottom. The view was fabulous. I could see Townsend Valley clearly both by day and by night. Colonel Return Jonathan Meigs started the Meigs Line from this very mountain in 1802. The Meigs Line retraced the last nine miles of the Hawkins Line from what is now known as Meigs Mountain to the crest of the Smokies at Mount Collins. There the Hawkins Line terminated. From Mount Collins, nine miles ahead, Meigs followed a different line to Brevard, North Carolina. *The next nine miles—because of the remoteness, the steep jungle-like terrain, the impenetrable vegetation that claws at clothes and skin—will be the roughest part of my trip. It grabs your hat and slings it back 30 feet. It's loaded with "rhododendron hells" that go for miles. You can walk out on rocks thinking your feet are on the ground when they are actually several feet above. It is hard to crawl under it, over it or around it. Sometimes you just have to get on top of it. There is a dust on the leaves that, when disturbed, fills the air producing coughing. I hope I won't cross a yellow jackets' nest because it is impossible to run. I have a feeling I will take several of these two-day breaks. I am down to averaging only three miles a day, and I fear my energy has left me.*

After the two-day layover I started for Blanket Mountain. I dropped down off Meigs Mountain and followed the compass to the left fork of Marks Creek. The valley was quite wide, and there were remains of old home sites. Near these sites grew the ever-present apple and walnut trees. At Marks Cove there was a large boulder under which the Cherokee camped while hunting in the area. It was called the Bald-Headed-Man Rock. As many as 15 Cherokee had lived under the rock. Arrowheads and flint were everywhere, indicating several years of occupation. The Cherokee had fled to this area and several others in 1838 to escape removal by General Winfield Scott to Oklahoma on the Trail of Tears. There were a total of 600 Indians living in the mountains—half of them starved to death.

I recalled driving up to this area by jeep early in my career. Now that road was closed because the bridge had been washed out on the Lynn Camp Prong.

I crossed Marks Creek at elevation 3240 feet. I began a most arduous climb up a slope of loose, granite-like rock. The rocks were medium size and would teeter back and forth precariously as I ascended. After a difficult climb of several hours, my ankles hurt and my legs were sore. Many times I fell down, scraping my hands on the rocks.

The valley seemed to be enclosed very well with only one exit and entrance at the bottom—the southwest end. It was protected from high winds and storms. I finally topped out at a clear spring issuing from under a large stone. *This is where I will camp!* Despite a full day's hike I made only two miles' progress. *I'm afraid my energy is going away.* The wind howled around my tent that night. My aches and pains set in and gave me no relief. After breakfast I prowled the 4608-foot peak Meigs had named "Blanket Mountain" on a survey when he came through here five years after Hawkins' survey. I found what appeared to be a large boulder overlook. I believed that the large stone would have been where Meigs laid his infamous red blanket. This was a marker that Meigs had placed to back-sight from Bent Arm and later from Meigs Post on Mount Collins.

The blanket would have been affixed in some way on the face of the rock. Another way to have secured it would have been to tie it high up between two trees. Whatever the method, I could surely understand the difficulty they faced in identifying the exact spot on top of the Smoky. Every mountaintop in this view looks similar with the exception of the deep dip created by Collins Gap. So I believe Meigs tied down the red blanket, then went quickly to the crest of old Smoky at Mount Collins. He then would have walked along the mountaintop, sighting back to the blanket. Eventually, his compass would have revealed to him the point at which he was back-sighting along S76°E.

Perhaps Hawkins wanted Meigs to be sure his survey had been correct. This could be why Meigs reran the last nine miles. I wonder if the point marking the terminus of Hawkins' survey and the beginning of Meigs' survey was originally named Hawkins Post before it was called Meigs Post in 1802. Had Hawkins worried that his post was at the wrong location? When Hawkins Line ended in 1797, immediately, the Pickens Line continued to the area near Tryon, North Carolina. This placed the white people of the Hendersonville area behind the Cherokee boundary. Hawkins had been called to Knoxville by President Adams to arrest Governor

Blount. The governor of the territory south of the Ohio had been charged with sedition. History tells us that in 1799, Colonel Butler ran a survey line bearing S27°E from Hawkins Post. The line went from Hawkins Post to a point where the Little River joins the Oconee River in South Carolina. Three years later, Meigs ran a line S52°30'E from Hawkins/Meigs Post to a point near Brevard, North Carolina. I believe the Butler Line had been rejected by the Cherokee; hence, the survey moved farther north toward Brevard. *So several days ahead on my journey I will come to this Hawkins/ Meigs Post.* The post, as the ending and starting point for several important surveys early in American history, remains a reference point to this day. It was, among other things, the key survey point for the Smoky Mountain National Park boundary.

My time had been well spent on Blanket Mountain. The two days' rest had helped me recover.

***

The next morning I set my staff into the soil. The compass led me down the hillside through laurel, maple and oak. I came to a small stream flowing toward Jake Creek. Along its bank were wild garlic ramps. I gathered several for the noon meal. At twelve noon I crossed Jake Creek Trail and walked into Backcountry Camp #27—elevation 3480 feet. Several young folks were there, and I shared my ramp-seasoned Ramen noodle soup with them.

They were surprised that when I prepared to leave, I wasn't taking the trail. I assured them I knew where I was going! *Do I know where my journey takes me?* I walked across the high, wide valley of Jakes Creek. Soon I came to a ridge. I found no blazes on the ridge, so I continued to the right fork of Knewt Prong—elevation 3680 feet—which flowed in from my right and exited to my left toward Elkmont Campground. The creek flows down a rather steep, high valley.

I noticed that an old man-way traversed the stream coming up from Jakes Creek. Man-ways were unimproved paths—not officially trails—maintained through the perpetual usage by large animals and Native Americans following paths of least resistance. It climbed to an old gant lot on Miry Ridge. Gant lots were gathering places where, from late summer to early fall, herders would hold the cattle of different owners. The owners then would come claim their cattle in November and drive them back to Elkmont. If an owner was late coming to claim his stock, the cattle would become "gaunt" or lose weight, staying in the small lot. *I wish I could remember whose gant lot this was. The name may be Henderson's Gant Lot.* The spring at the old gant lot was the very beginning for the right fork of Knewt Prong. At this lot there were remains of a herder's cabin and an old fenced area of pasture.

Near there was Jasper Mellinger Death Ridge. I recalled that in 1904, Mellinger was en route to the Everett Mine on Sugar Fork to work as a blacksmith. While traveling up Miry Ridge, the fog became so thick he stepped into a huge bear trap. Days later the trappers returned to find Mellinger barely alive. Fearing the law, they killed Mellinger and dumped him over a bluff. They then took their trap and left. Tradition says that one of the killers was haunted by what he had done for the rest of his life.

I could find no indication of the Hawkins survey on the right fork of Knewt Prong. I gathered my energy to climb the ridge toward the left fork of Knewt Prong. *The sidehilling is taking its toll on my stamina.* Once over the hog-back ridge—shaped like a boar's hump—I dropped down into the left fork of Knewt Prong. The creek was choked with rhododendron and blow-down. I climbed out by going directly uphill toward Bent Arm Ridge. *The fun and excitement of my trip seems to be waning for the moment.*

At elevation 4200 feet, I came to a pretty, horseshoe-shaped valley. I walked up the valley a quarter mile and came to the right side of the horseshoe at elevation 4400 feet. I observed edible morels in the valley and lots of white phacelia as I neared Bent Arm. At elevation 4600 feet I came to a wide gap on the ridge. An old "government trail" was evident on the crest of the ridge. This one was a result of abandonment by the Park Service of a formerly official trail. To my left I could see the excellent rockwork once maintained by the Civilian Conservation Corps—the CCC boys. The CCC was once a way of providing employment for out-of-work men. They gained skills in the process, for use in later careers, such as masonry for later bridge building.

Nearby was a fairly large tree. Large trees were not uncommon in the area, but what was inside the tree was uncommon. Inside the live tree was a block of white quartz—perfectly square. It looked to weigh eight pounds or more. *Did Hawkins cut out a square in the tree and place the rock in it to indicate the survey line?* Standing with the tree to my back, the reverse reading pointed to Blanket Mountain! The forward reading pointed toward the Hawkins/Meigs Post. It is understandable that the team left a special quartz-rock marker because Bent Arm had strategic importance. From previous experience, I knew the Cherokee would place white quartz in trees to indicate their routes. On the tree were numerous human carvings that could warrant later investigation. I stood in front of the tree and thought about the surveys. *I believe this could be a line tree constructed by the Hawkins/Pickens survey with the help of the Cherokee.*

All around me were the telltale signs of past grazing activity. The native mountain rye grass grows wild where a previous cattle grazing has occurred. The cattle were kept from straying by salt blocks housed in chestnut logs, tended by a summer drover. There still remained a sampling of the mountain grasses that grew in these open areas long ago. I pushed onward down the branch called Hostility. Later I rock-hopped across the creek called Battle and finally rested for the night in War Hollow—at elevation 3000 feet. As I crossed the Fish Camp Prong of Little River, I recalled that one landowner sold his land to Little River Lumber Company by the boundaries set by these surveys. I tried to rest my weary body, but the location was appropriately named. My own body did war, battle and wax hostile toward me.

The next morning I fought back the urge to stay for several days' rest. I pushed on, traveling up a slippery slick hollow to Little Goshen Knob. *Goshen means "plenty" in the Bible. The land, Goshen, was the promised land of Canaan.* I saw plenty of trees, rocks and dirt.

Approximately four miles south was the location of a gold mine. An individual discovered it near the Appalachian Trail in the 1920s. A local gold rush occurred, but, sadly, they found one ton of ore would yield only $1.27. It would take all week to hammer out that $1.27! A similar mine was found at what is now the Elkmont section of the park; there again, the content was minimal.

From Little Goshen, I sidehilled across a mountain and crossed Split Branch. Then came a very bad climb over a high mountain— elevation 4100 feet. After a dreadfully slow ascent, I could find no sign of old blazes on the mountain. I sidehilled for about a mile and crossed Striped Hollow. After going one-half mile further, I could hear the roar of a large river below.

LOST

Late into the evening, I came to Little River. The stream flows from the crest of the Smokies in a northerly direction. The recent rains had swollen the stream above its banks. I searched up and down the stream until I found the best possible crossing. Retrieving my climbing rope from my pack, I tied a good-sized rock on one end. With my right hand, I spun the rock clockwise and let go. The rock spiraled over a large oak limb on the other side of the stream. The rock-bearing rope had successfully wrapped itself several times around the limb. I pulled hard on it to test it. On my side of the creek, I looped my section around a stout tree and tied it with a slipknot. As I started across the raging water, I fed out the remaining rope. Midway across, the river floated my feet off the stream bed. While I held tightly to the rope, my right foot again settled to the rock bottom. The force of the water wedged my boot between two rocks. I couldn't get my foot loose, and my crossing-rope was now stretching, sagging lower and lower. The water was now pushing against my chest as I sank closer and closer to being fully underwater.

If I let go of the rope the river will bend me backwards and submerge me with my back under the water. I will not be able to right myself. I must take drastic measures. With the waves lapping at my chin, I reached with my right hand and pulled out my long knife from my waist belt. Directing the knife underwater, I found the spot where my ankle joins my leg. I always kept the knife razor-sharp, and I knew it would cut through almost anything.

The sweat popped out on my forehead as adrenalin flowed. I felt the power surge to my muscles. With one deep thrust and slash, out popped my leg—minus the right boot. After several tries, I found the boot was irretrievable. Thankfully, my training took over as I conducted a Tyrolean traverse to get to the other bank—hand-walking along the rope facing upstream and moving from right to left. Climbing out on the other side, I pulled hard on the end of the rope. The rope untied and fell into the river. I retrieved the rope and hung it to dry as best it could by making large loops over my shoulder. *What a mess! I have one boot on my left foot and no boot on my right foot.* I removed the left boot and secured it to the outside of my pack. My camp sandals would have to do until I reached my next camp.

Climbing up the riverbank, I came to where Meigs Post Branch joins Little River at elevation 3280 feet. I walked approximately one mile up the branch to where Sweet Creek comes in from the left. There I camped for the night at an old logging operation— elevation 3720 feet. The area was fairly flat with a lot of rhododendron. I found an old wood stove and cleaned out the old clinkers. Inside the stove were several old and new rats' nests. I gathered these under my shelter for later use. A quick survey of the forest yielded some dry pine knots high up on a tree. I combined the extremely dry rats' nests with the pine knots and had a nice fire going in the wood stove. As I warmed myself, I cooked my meal of more Ramen noodles and ramps. I quickly built my shelter as a cold, wet rain began to fall.

After hanging my wet clothing above the wood stove to dry, I retired to my tent. I crawled into my bag and slept the sleep of discomfort throughout the night.

In the wee hours of the morning, I was awakened by a sound— a familiar sound! Bear. *A bear is in my camp!* I couldn't believe how quickly I came to life. Maybe I was just tired of the Smoky Mountains pushing me around. Out of the tent I emerged with two hiking sticks in hand, ready to give that bear the whipping of his life. There he was—15 feet off from the fire. *What's that he has in his mouth? My pants! My pants! He is eating the right rear pocket of my pants. This is too much!*

With sticks flailing wildly, I attacked. "Drop my britches, Mister Bear . . . or suffer the consequences!" The bear did heed my advice and let go of my clothes. I retrieved my britches and observed a hole where the right rear pocket used to be. With that incident, I brought my half-dry, wood-smoked clothes into the tent.

The next morning the rain was even heavier. It had rained all night. Everything I had was wet and getting wetter. What made me look at my watch, I'll never know, but I did. A cold, desperate feeling came over me. The date on my watch was June 4. *June 4! I'm late! I'm behind schedule.* I was four miles from my destination, and the elevation gain ahead of me was 2200 feet. *My sweet Julia will be at Collins Gap June 6. Can I make it to Meigs Post within the next two days? I must not panic! Take one step at a time . . . but I have noticed my steps are getting shorter every day. Will she worry about me? If only I had a radio I could call.*

I spent the morning constructing a replacement boot, using my right sandal, a hunk of leather and heavy-duty fishing line. The boot was not fancy like the one left in the stream, but it would do.

With one step at a time, I made my way over fallen logs, brush, briars and brambles. First of all, I would try to make it to Wilson Falls and rest there. The makeshift shoe started failing me after about two hours. I didn't have the ability to lace up the shoe to strengthen my ankle. Every other step was a wrenching, twisting experience for my right foot. The more I shifted my weight off the right foot, the more it burdened my left foot and my two hiking sticks. While I was crossing a large jumble of fallen trees blocking the creek, my right ankle twisted grotesquely. I suddenly fell and just lay there quietly for what seemed like an eternity.

I finally dragged myself to a standing position and limped over to some young saplings. With my Arkansas long knife, I split one of them and wrapped it with my old shirt, creating a splint, which I then laid along my right ankle. I tightened the splint with my belt, and it immediately felt a little better. I struggled slowly up the stream, dragging my ailing limb.

Late in the evening, I came to what looked like Wilson Falls—more like a series of small falls than one single waterfall. My chains told me I had come 6400 feet up the Meigs Post Branch—only about one mile and one quarter of walking, but in my condition it took me all day.

I must look at the situation positively, or I will lose faith in myself. Have I not been in much worse situations than this? Yes, I will dig down deep to that inner strength. I will beat these difficulties. I will make a comfortable camp, and I will heal myself and move on.

My elevation at Wilson Falls was 5200 feet. I had found no blazes between Bent Arm and this location. I began to believe that they did not exist! They are not here! The line must have been shot through the air—right over where I was now located—but never laid on the ground!

My camp was on a southeast-facing hill—high and dry. I started work on my shelter. I believed I would be staying at Wilson Falls until I could heal my ailments. The first step was to gather lots of dry leaves, debris and different lengths of poles. I stripped a fallen linden tree of its rope bark.

I located a standing tree with a large fork ten feet up on it. I placed a twenty-foot-long ridge pole in the fork and tied it with linden rope. The other end of the ridge pole rested on the ground. I drove stakes on either side to secure it and then placed ten-foot rafter poles twelve inches apart along one side of the ridge pole. The process was repeated on the other side of the ridge pole. I secured these rafter poles to the ridge pole with linden rope and then tied twenty-foot saplings cross-ways along the rafter poles, six inches apart on both sides. I filled the empty squares with dry leaves and other debris and added more saplings and layers of debris until I had a four-foot-deep insulated structure.

Next, I peeled ten-foot lengths of poplar bark to serve as shingles over the structure. I then placed two eleven-foot sections of poplar bark on the top to act as a ridge cap. I built a separate roof slightly in front of the hut. This was made from poles and slabs of poplar bark and was quite high to allow for my fire pit under it. Finally, I constructed a thick door out of saplings and leaves to close my hut entrance at night. The remaining leaves and debris I spread inside my hut floor. To create the fire pit area, I hauled many stones and placed them in an ever-widening circle on the earthen floor in front of the hut. Next, I gathered lots of dry wood, wet wood, green wood and many rats' nests and pine knots. I built up a small-size fire to warm my food and heat my medicine stones. They would work like a hot-water bottle or heating pad. Over my fire, I heated a stout soup made from wild ramps and rock tripe. The latter is a lichen that grows only on thunderhead sandstone in the Great Smokies. Undisturbed by children, in the wilderness it will often continue to grow. The larger ones on the rocks may be as many as 3,000 years old. I knew the rock tripe had antibiotic qualities that might benefit me.

After supper I banked the fire with wet wood and green wood. I then made a poultice from juniper. This would reduce the swelling of my ankle. I made two additional poultices, one from witch hazel and sassafras root bark, and the other from poplar bark, white pine bark and sphagnum moss. After using all those treatments, I would know if I had a sprain, strain or a more serious problem by the degree of reduction of the pain and swelling.

At bedtime I prepared a bowl of hot water and boneset solution, made from the plant by the same name that was growing in wet areas. I tugged, rolled and otherwise maneuvered a collection of hot stones to form a circle around my injured limb. I soaked the hot rocks with the boneset solution and placed juniper boughs over the entire traumatized area. My ankle lay in a cradle of moist, hot herb-layered rocks.

I was pleasantly surprised to observe that the pain had subsided a little when I arose the next morning.

The next day I was to rendezvous with Sweet Julia. When I tried to walk on my ankle, it seemed that I would require more healing time. This was my fourth day at Wilson Falls. Overhead, I thought I heard the sound of a light aircraft making a tight turn over Wilson Falls. Trail runners would be on the marked park trails by now. Trail runners' sole priority was to stay on assigned trails, looking for sign, proceeding as fast as possible.

They would not find me there. I was walking a two-century-old survey line. It was June 5, and darkness was overtaking Collins Gap and the high places of the Great Smokies. *Is it real or is my mind playing tricks on me?* My leg hurt. I feared that it might be broken after all. It would be dark soon. *Sure would like to have seen an aircraft by now. Sure do hope that they know I'm in this area. Surely Dwight will figure this out, knowing my plan to follow the Hawkins Line. He is probably on it right now . . . probably got on my trail somewhere about the Blowdown on Thunderhead. Boy, what a place that must have been when Bill Walker first saw it . . . probably just after The War Between the States, shortly after he and Miss Nancy moved into the Middle Prong in 1859 . . . wonder if he knew about the Hawkins Line . . . perhaps his land papers had reference to the call—S76°E. Shucks, I'll bet he didn't even have any land papers, especially not a registered deed. It's a real shame how he was paid by the lumber company for only about one-tenth of what he felt he had laid claim to. All of Thunderhead Prong . . . John's New World, as Nancy called it . . . even the lumber company named their camp the New World when they began cutting after ole Bill died . . . and had been convinced to sell that beautiful country to them . . . a real shame . . . ole Bill tried so hard to save those big ole poplar trees up there from the crosscut saws and skidders. Yeah, the Blowdown . . . that may be where Dwight is right now. He always could find tracks, even when I tried to ghost him and not leave any sign at all . . . like the time near Cumberland Gap when Daniel Boone walked one evening for several miles, thinking he had hid all sign from anybody. He felt he had done that very well. Too well! He got a strange feeling about it all. "No sign of no one behind me," he had thought to himself. "Too good. Much too good." He turned around quietly and looked back into the twilight woods. No one. He couldn't stand it. He then whispered gently, "That you, Simon?"*

"Yep," replied the voice of Simon Kentron from somewhere back in the twilight.

Yeah. Yeah, Dwight would do just like that. He would find me. Yeah, he's somewhere in on the Blowdown . . . or is he? Wait a minute, he would know that I wanted most of all to find the line between Blanket Mountain and Meigs Post! He would know that I would be in here somewhere by now. By golly, he's several days ahead of that. He can't be too far from here now! That you, Dwight . . . that you, Dwight?

Gosh, no answer. Not unless he's toying with me. No, he would not do that! He knows by now that I'm hurt. He will think of Wilson Falls. He and I talked a lot about how J. O. Morrell's father knew Mr. Wilson, who testified in the case where the lumber company was accused of moving Meigs Post to the west near Siler's Bald. Yes! Yes, he will think of these falls—Wilson Falls! Good luck, Dwight!

I had tried the last of my poultice-hot-rock remedy. The poplar-bark and white-pine-bark-and-sphagnum poultices seemed to have worked. *I must thank my old friend Chief Birdtown for teaching me these ancient Cherokee remedies. I feel much better. The swelling has gone away and I can walk better. But I must address the loss of my right boot. I must stabilize the movement of my right ankle or I'll be laid up again.*

***

Sweet Julia had waited anxiously at Collins Gap after parking there around noon. Vinn had said that he would come out at the gap about midday on the 6th. As a seasoned, experienced ranger's wife, she knew that estimated times for completing long distance walks through the mountains should be very flexible, especially with such a venture as Vinn's, which involved so many days. *He will be here,* she thought to herself. *He always planned these things so very well. So very well—to a fault! He*

*never wanted me to worry about him—he will be okay . . .
"After all," he would say, "I'm being paid to know these things."*

She departed Collins Gap about two hours past dark. For her
not to worry was like telling the wind not to blow. A ranger's
wife spends a lifetime worrying! So many times their husbands
would leave out in the morning, planning to return at the end
of the day—only to show back up some three or four days later
without the means to let their wives know not to keep supper
warm that night . . . or the next night . . . or the next . . . Vinn
and numerous other rangers had no idea of the anxiety during
those hours their wives spent alone at an isolated ranger station
awaiting their return.

Sweet Julia drove back to Collins Gap on the second day and
walked to Meigs Post. The third day she returned and walked up
to Meigs Post and back three times. She paced back and forth
on the trail on the off chance that Winn might have shortcutted
to the car. On the fourth day—June 10—she decided that she
could not allow the perpetual twilight of the Great Smokies
to enclose the mountain tops without saying to herself, *That's
enough. That's it! I must tell someone! Yes, Dwight and R.J.
Buckingham! They will know best how to get some help, and,
yes, our three strong sons! They would want to help find their
dad. I know that's how Vinn would want it. My Lord, please
help him! Please let him not be hurt!*

Upon arriving back at Tremont Ranger Station, Julia dialed
Dwight, but his eight-party phone line was busy.

She drove to Middleton's Store in Townsend where she knew
that Dwight sometimes stopped in the evening to buy some
herbs. There she found him and stressfully explained the saga of
the last few days. She attempted to be professional and factual,
but Dwight knew she was worried. He was also well aware
of Vinn's obsession to follow the Hawkins Line and all that it
would entail.

Dwight assured Julia that the best thing to do was to contact Chief Ranger Peter Moss as soon as possible. He was certain that Pete would help with the search. "Julia, I will contact R.J. Buckingham and Chief Birdtown in Cherokee. R.J. knows this country as well as I do, and the Chief has helped all of us learn about finding folks in these mountains. They are both excellent trackers as well. Go ahead and call up your sons. They have good mountain savvy, which they learned from their dad. Didn't you tell me that Vinn had nicknames for each of them?"

"Yes," Julia replied, "he called them Kingfish, Shine Boy and Scat Man."

"Wow," Dwight said, "I'll have to follow-up some day on how they acquired those names. Get in touch with them and see if they can be here first thing in the morning."

***

"Search is an Emergency," Chief Moss said! *Where have I heard that before? From old Vinn of course!* I had joined other searchers at the training room at park headquarters. Among them were R.J. Buckingham, Chief Birdtown and Vinn's three strong sons. Sweet Julia and the Chief Ranger's wife, Lucille, had been there since before dawn. It was reassuring to see the Whitehead brothers present. Their tracking skills and woods savvy were most welcome on this occasion.

"Dwight, I waited three days at Collins and walked the trail up to Meigs Post several times. It's just not like him! I fear he is hurt . . . or worse."

"I know, Miss Julia, but I really believe that Vinn is okay. He's probably taking longer just because he really wants to find traces of that old survey having actually been marked and blazed, especially through the section he's probably in right now." Beneath my calm expression, however, were serious concerns for my old friend.

Chief Ranger Moss began to put things in motion. "Men, we will divide up into teams of trail runners. Dwight, your group will go up Jakes Creek and cut sign. Bud, take your squad and check Panther Creek. Grady, you and Arthur lead your group up to the top of Sugarland Mountain. Lennie, you go up Miry Ridge Trail. If anybody finds credible sign, be sure and flag it well. Dwight's group will look at it before we start checking anything off trail. We will be calling in tracking dogs shortly to assist in the search.

"Ladies and Gentlemen, most likely Vinn is okay. We have made copies of the diary and route map he gave Julia when she last contacted him on Middle Prong. They may be a key to finding him. Be sure to take plenty of 'D' batteries for your radios and flashlights. We will use radio channel 'B' for search activity, so be sure to use the main channel only if you have found something. Are there any questions?"

"Yes, sir," I said, "would you mind if we said a prayer for Vinn and the searchers?"

Chief Ranger Moss bowed his head and said, "Dear Lord, please watch over the lost person and these young men and women and bring them all back to us safely. Amen!"

***

Folks in my group included R.J. Buckingham, Chief Birdtown and Vinn's three strong sons. This was quite a combination of seasoned experience and the strength of youth. All of Vinn's sons had provided blessed assistance to him during dozens of traumatic incidents—the type that come to the door of an isolated ranger station, requiring immediate response in the form of searching, putting out fires, bandaging, broken-bone-splinting, and helping others deal with a variety of emotional shocks. The type that come when the nearest additional help is miles away. R.J. was familiar with every mile of the dark wilderness of Thunderhead Mountain, the Middle Prong of Little River and the Blowdown—the site that suffered a great impact from the 1936 hurricane that blew vast amounts of the forest down. To a remarkable degree, R.J. possessed that almost uncanny power that few, other than the Indians, have ever acquired—reading the signs of man or beast in the woods. And among those Indians would also be recorded the name of Chief Birdtown. Before the dawn of this day, the Chief had slung across his shoulders his ration of corn dodger, streaked middling and stone-ground coffee—all rolled up in a home-woven blanket, tied at the ends with leather thongs. He had tightened his leather belt, picked up his dogwood tracking stick and offered a blessing in the native Cherokee.

Searchers and a group of trail runners drove to the Jakes Creek trailhead. This was about four miles from Blanket Mountain. The darkness was still lingering as we departed up the trail. We all walked along the edge of the path so as not to spoil any sign. We could see no tracks that remotely measured 19 inches from heel to toe. Some of our team occasionally spotted possibilities. *How you can turn a bear's track into Vinn Caroon is beyond belief!*

Then somebody smelled a dead body. We responded . . . to find it was skunk cabbage, which I must admit smells just like decaying flesh. We became aware that Kingfish, one of Vinn's sons was there in the group contemplating the skunk cabbage.

"Gentlemen, my dad is not dead. Don't look for him among the dead. Search for him among the living." With that, the tall young man quietly walked away. I made a mental note not to look at any more skunk cabbage, and I sure hoped we didn't come up on any vines of the cadaver plant.

At Campsite #27 we observed three tents and several people around a campfire. The offer of hot coffee was welcome. *Why is it that when a park ranger arrives anywhere, someone will have a question for him?*

"Ranger, when do the Rosy-Ding-Dong's bloom?"

"No, we'll be working our summer jobs. Maybe next year."

I've found it easier to go along with the visitor's mispronunciations than to try to correct them. "Mid-July, sir. It's not far off. Will you be able to return to view them?"

Out of the corner of my eye, I saw something familiar. It was just at the edge of the glow from the fire. *A message left by Vinn!* It was a rock leaning against a poplar tree. The rock was tapered, the apex pointing toward Bent Arm Ridge. *The old rascal! It would make sense to find something here because Hawkins Line crosses right through this campsite. He always leaves lots of insurance, and if his energy levels are low, he might leave extra signs.*

The trail runners were ready to go on up the trail. I bade them farewell on their assignment. Our groups would be going a different ways from here on. To my consternation, I was joined at the lead by Phil, a road-patrol ranger with minimal search experience.

We struck out toward the next rock leaning against another tree. Phil did not seem to trust my prediction that there would be another rock pointing toward Bent Arm. There it was an hour later. I'm certain Phil was just about to ask, "Where's the next rock?" The flashlight reflected off a tapered rock ahead. Phil was quiet through the discovery of the next three rocks.

Phil was older than me chronologically and had worked as a ranger longer, but he had yet to realize his full potential when it came to understanding God's nature.

So, we continued late into the night following rocks into the wilderness. Vinn's trail led us across both forks of Knewt Prong. Pretty soon Phil accepted the notion that I must know the answer to the "Mystery of Leaning Rocks." The darkest hour arrived just before dawn and along with it a cold wind. I bunched my collar and shrunk my neck into my shoulders.

"Dwight, where's that cold wind coming from?" Phil asked.

I licked my finger and started a slow circle holding my finger in front of me. As the finger passed the direction of Elkmont, a chill came to the wet area of the finger. The chill left when I turned past Elkmont. "Yon cold wind doth blow from Elkmont, lad."

We climbed out on top of Bent Arm as a chorus of awakening birds greeted us. I grasped the radio, "Chief, this is Dwight!"

"This is Chief."

"Chief, we've followed Vinn out of Jakes Creek to Bent Arm Ridge. His sign indicates he is walking slower than normal but he does not wander. We are able to track him in a quicker fashion, in that daylight has arrived. I suggest that trail runners give Fish Camp Prong and Three Forks a check."

"Roger! Out."

Phil was staring at me. "How did you know he was walking slow?"

"I had noted the shorter stride indicated by the footprints."

"Footprints?"

"In the leaves, Phil."

Phil trudged quietly.

"Did you know smokers have a shorter stride than nonsmokers?" I asked. Vinn didn't smoke, of course, but I was feeling a need to encourage Phil with some teaching. "Another thing about stride, you can determine the time of day by comparing day stride with night stride. You can tell when night overtook someone. Sign is easier to track at night when laid down."

"Why?"

"The walker stumbles more and bumps into things more."

Phil walked on, deep in thought.

"Group, grab onto yourselves. We are going to track at a faster pace. Just stay with me—okay? And watch your footing." Down the side of the hill toward Fish Camp Prong we descended, nearly in a dead run. Plenty of sign was visible along with the occasional rock against a tree. The sign led us to the edge of the river an hour later. We never stopped to hop rocks but instead, splashed waist deep into the stream and emerged on the other side. The sign crossed the trail and came to a camp on the side of Goshen ridge. *Vinn's camp! But it's old! Way old!*

I realized that I must tarry a while to look for Vinn's new marker. At first I had missed it as my eyes swept the area heading toward Little River. When I swept my eyes back across the area, there it was! *Artist Fungi! Artist Fungi!* Leaning against the tree facing me was white Artist Fungi.

Artist Fungi grow on locust trees. Some are as high as eight to ten feet on the tree. The bottom sides are brilliant white. The tops are bronze-looking. They look like three-quarters of a pancake sticking out of a tree. Rangers will tear them off, and, with their trusty U.S. Government pens, they can write a message on the white side. The fungus is leaned against the base of the tree with the white side out. When the finder stands before the tree to read the message, he or she is facing the direction to proceed next.

We quickly walked to it. A message was attached with an arrow on the brilliant white part.

> Dwight, if you are reading this, then you must be on my trail. I am okay here, but the trip has taken it's toll on my energy. You take care, lad.
> —Vinn

I knew that now we had a pattern. "Fellas, do you know the markers we will see from here to Little River?"

"Yes, Dwight, I believe it will be Artist Fungi messages," Phil volunteered.

We again struck out at a dead run passing several more Artist Fungi markers. Over the last ridge, we could hear the roar of Little River. Following Vinn's track we came to its edge. We needed to shift to a creeping-crawl mode. We needed to be on our hands and knees with our eyes literally on the track. The sign had become very faint, and it would be easier to make a mistaken conclusion that Vinn did not cross here.

In creeping-crawl mode we proceeded up Little River, but Vinn's tracks were spotted way up on the river bank. We were walking farther downslope, perhaps ten feet closer to the river. Something had driven him up the bank. *It's a flood! Vinn is searching for a good crossing!* We proceeded upstream and came to an odd sight. There, right in the middle of Little River, was a huge boot stuck tight in the crack of a rock.

Oh, no, no, is he okay? Is he drowned? Focusing intently on the boot, I literally propelled myself to its location, leaping down the bank and landing adjacent to it. Standing in a foot and a half of water, I had started my inspection when my peripheral vision caught movement to my left. I looked back up the hill and saw R.J. and Chief Birdtown coming toward the river. I asked if they remembered the recent drowning on Little River. As we looked at the solitary boot, their eyes told me they remembered. When a tuber floating the river fell off and his ankle lodged in a crack of two rocks, the river had forced the tuber under water. Bystanders had tried to keep him above water, but they could not stand up. The man begged for them to sever his ankle and free him, but no one had the means or the intestinal fortitude. I looked at R.J. and whispered, "Vinn's boot has been cut clean through at the ankle."

Chief Birdtown and R.J. turned back on the trail and went to meet Vinn's sons to prepare them for what we had seen, although I had not yet freed the boot sufficiently to inspect the contents. These boys grew up assisting their dad in every kind of rescue in the book, but I could see they were concerned as they approached. I scanned down the river. No sign of Vinn. *If drowned, he would be miles downriver. But the boot was cut clean through, so he could be nearby, seriously injured or dead!* "Guys, Vinn was in trouble here. He had to use that Arkansas toothpick he always carries in his belt. He had to cut himself free."

We pried with our hiking sticks and finally recovered what proved to be the lower part of Vinn's right boot. It appeared that the laces had been purposely cut as well as the leather. I tied his boot around my neck after the relief set in that it was empty. It had changed from a nine-inch-high boot to an ankle-high. We went to the other side where Meigs Post Branch comes into Little River. To my great relief, there, plain as day, were Vinn's footprints, albeit a different pattern from the other side of the river, a left footprint and a right footprint. They were headed up Meigs Post Branch. "Phil, let's look for Vinn's next set of sign."

I nearly stepped on the marker. There, stuck in the ground, was a limb with a fork two feet above ground. Another limb was laid in the fork pointing toward Wilson Falls. We set off in a dead run through very rough country. All along the way we found the fork-limb markers. Hours later we came running into another one of Vinn's now-familiar camps. Footprints entered the camp and exited without veering right or left, telling us he had spent the night without deviating from the survey route. If he had left the trail, we would have needed hours to explore each path.

Not wasting time, I scanned upstream in the direction of Wilson Falls. There ahead of me was a broken limb. Based on the position of the footprints, the limb had been broken by Vinn's right hand and that pointed us in the direction he went—toward Wilson Falls. Again we set out at a dead run and sometimes we got a whiff of smoke! *Smoke! Ahead! Coming down the valley!*

I glanced back and was proud to see Phil determining the direction of the smoke source. He licked his finger and held it up, casting it from left to right. "A chill, a chill!" he claimed, pointing.

"You're lookin' straight up the holler." R.J. quickened his pace. "That's Wilson Falls!"

"That's where he is!" I called.

"Dwight! Dwight! Come In!" a voice called over the radio.

"This is Dwight!"

"Dwight, this is Julia. We are with Civil Air Patrol Colonel Benton flying over Wilson Falls. We think we have spotted Vinn's smoke. What is your location, Dwight?"

"I am at Meigs Post Branch where it joins Sweet Branch, Julia, and I have found Vinn's right boot. Julia, I am probably less than one hour from Wilson Falls. We are headed toward him fast, Julia. Over and out."

We could now see the plane we had been hearing. The sound of the engines changed as we watched the Cessna 172 make a tight left-hand turn over Vinn's camp ahead.

The closer we came to the camp, the farther Vinn's sons and I got ahead of the group.

WILSON FALLS

A voice! A voice! Someone using a megaphone from a plane. But that's not just any voice. Julia! It's Julia talking to me in a relaxed, even and steady voice.

"Vinn! Vinn! This is Julia. If you are at Wilson Falls, give us a signal. We will circle here for a while, but not for too long as I'm about to lose my breakfast!"

I threw some green leaves on my fire, and the extremely bright white smoke drifted upward into the sky. Julia came back on the loudspeaker.

I was amazed how calm, cool and organized she was. But that came from years of being a good, reliable teacher who genuinely cares for her students while keeping them on task.

"Vinn! Vinn! Dwight is on your track. He has just found your boot in the stream. He should be at your location in one hour or less. Vinn! Vinn! We are tossing a large message dropper with a bright orange flagging attached to it. Read the enclosed cards and wave the enclosed colored flags that match your needs. If everything is okay, don't wave anything."

I saw the message dropper fall near my camp. I retrieved the packet and stood looking at the plane, not waving anything. The voice came again as if from heaven.

"Vinn! Vinn! Get under a tree. A package will be arriving via air-mail. I'm dropping you a gift in a small orange bag. We will see you at the top!"

Down through the canopy came the orange bag, but it hung up in a large birch tree. The plane banked sharply and flew toward Sevier County Airport. Since I was not in any shape to climb a tree, I busied myself gathering my gear and putting out my fire. My camp had been a comfortable one. I would miss this area. *Surely God's hand created this wonderland,* I thought. *The forest is formidable, but God has also placed the remedy for one's difficulties there—often in close proximity—unless man has drastically intervened.*

Suddenly the hair on the back of my neck stood up. I reached for my sticks. *Something walks through the forest.* "Is that you, Dwight?" I looked toward the low hill where the sound came from. All I saw were three tall trees moving toward me. *Wait a minute, trees don't move! Those are three tall men I seem to know. It's my sons! They are coming for me.* The moment overwhelmed me. Before I called out to them, a slight silence was filled with my prayer of thanks.

My boys walked right up to me. Kingfish, Shine Boy and Scat Man! *Julie and I are so proud of all of them! Thank you, Jesus, for this blessed moment. Yes, I am being restored by faith, family and friends.* All three wanted to see my right foot! "That you, Vinn? That you, Vinn?" Dwight, R.J., Chief Birdtown and Phil came trailing into the camp and was a sight for sore eyes.

"Where you been, lad? I been looking for you."

"Well, Sir, it took me a while to pry this here #16 boot out of that rock, and it do weigh a ton, wet and all like it is."

"Gimmee that boot, lad. I'm tired of walking in my boot made of bark. Lad, you see that tree down there with the orange bag 100 feet up in it. See if you can retrieve it. Oh, by the way, good to see you. I knew you would come."

Dwight climbed the tree in no time.

Ah!! To be young again. To be able to swing on grapevines across rivers. To be able to run full speed up a mountain and not stop at the top but continue just as fast down the other side.

"Here's your orange bag, sir. It sure smells good."

"Yeah, Julia threw it out of the plane." I opened up the bag to find a pleasure worth the whole trip. It was Julia's prized, old-timey stack cake for which she is legendary. The group and I ate ravenously and left only the crumbs for the birds. I retied the cut pieces of what was left of my right boot laces and shucked on that extremely beautiful boot. I was able to bear weight again, and the cake and water seemed to restore my strength.

"Well, lad, let's get out of this place. I've enjoyed it here, but I wouldn't want to live here."

* * *

Vinn led off in front of me. I could tell he had been through an ordeal by his slower gait. He would tell me about it when he was ready. But there was an unspoken savvy between us. It's one of those friendships that are hard to find. I swear I've seen some rough country before, but this takes the cake. There were fallen trees everywhere. The balsam were dying here and there, and the forest was awash with young spruce invaders. We had only three-quarters of a mile to go but the elevation change would be close to 1000 feet—and all uphill.

We traveled slowly up the mountain, taking ample breaks and swapping tall tales. The five of us were joined by the remainder of our party—Phil, Chief Birdtown and R.J. By now the search had been called off. This might be the shortest search in history, but I'm glad we jumped on it immediately. Vinn obviously did all the right things to insure his survival. Often the extra time taken to doctor one's injuries pays off. As we approached the last half-mile, one of Vinn's hiking staffs broke. I saw the flash of that Arkansas toothpick come from Vinn's holster as he eyed a young sapling.

"Hold it, Vinn. Let's see if yon metal pipe growing out of that spruce will suffice."

"Metal pipe, where is it?"

"There, about 100 feet ahead." We ambled up to the metal pipe, and lo and behold, it was the handle of a Radio Flyer red wagon growing out of the fork of a spruce tree. "Vinn, this wagon has been here a long time. See, the paint has rusted away." I could see that the handle on the front of the wagon would make Vinn an excellent hiking stick.

Phil, as it turned out, brought everything but the kitchen sink when he hit the woods. We used his crescent wrench to take the bolt from the wagon. The handle came off easily. Vinn's number-two son salvaged a stout section of Vinn's broken stick and jammed and hammered it up the hollow wagon handle. With a wood screw provided from Phil's pack, he locked the wood onto the wagon handle.

"See here, Vinn, you even have a ready-made handle on the end."

Vinn's number-two son had the unique ability to analyze and fix anything. As we strolled onward up the hill about 200 feet, we noted Phil was not with us. He was still looking at the wagon.

Phil shouted, "Wait a minute, you guys. Where did the wagon come from, for God's sake?"

"Oh, I thought you knew, Phil! It came from the sky," replied Chief Birdtown.

Phil was quiet. I was tempted to tell him how factual Chief's supposed taunt was. In the Smokies, not reachable by boat or air, only by foot, LeConte Lodge flew in supplies from Gatlinburg-Pittman High School. Years earlier, they had to use Collins Gap, not too far from Meigs Post. One day, while they were air-lifting supplies, an item that was part of the sling-load broke and this wagon fell and landed in the spruce tree. It was a red wagon intended for Nathan, the son of the caretaker. I found myself reminiscing about Nathan. Oftentimes, I took time out of my duties to play bulldozer and sandbox with him near LeConte Lodge. On my many trips off the mountain, Nathan would be one of the first kids I'd see. He had no one to play with—his little sister was too young. He would say, "Ranger Dwight, come play with me." I would take my ranger hat off and get in the sandbox. Visitors passing by were perplexed at the full-uniform ranger minus Stetson, playing in the sandbox. Nathan was my buddy.

"C'mon, Dwight! We got a line to follow!"

We kept walking and were soon out of sight of Phil. We could hear him coming, sucking in great gobs of fresh air and puffing like a freight train. The mountain was getting a little flatter, and occasionally we could hear two whistles blow. The closer we got to Meigs Post the louder the whistles became. Finally, we could make out two ladies standing in what appeared to be woods similar to ours.

"They are at Meigs Post, Dwight. They are standing in the trail blowing whistles. Why, it's the chief's wife and Julia!"

We broke out of the forest and went directly to them.

* * *

It was overwhelming to see my wonderful wife! With my three sons, we hugged in a family circle, which will always be unbroken. *Thank you, Jesus!*

* * *

It's amazing we had not even perceived there was a trail at their location. We were 60 miles from Fort Southwest Point. This happens many lost people. They may walk within twenty feet of a maintained trail and not see it, only to turn and go back down the forest, thinking no trail exists there.

The ladies did take an interest in Vinn's wagon-tongue hiking stick. "What did you guys do? Take some kid's red wagon for a hiking stick?"

"God gave us that wagon, Julia. It fell from heaven ten years ago, we estimate. God placed it there for our needs," Vinn said.

"Vinn, don't confuse Phil any more! He may be at his limit of understanding," I cautioned.

Well, this was a fine evening with Vinn sitting next to Meigs Post, savoring the moment. I reflected on the occasion. Everyone has that one or two little things that they have thought of doing all their life. Vinn could chalk this one off the board and go on to the next one. Surrounding us were huge axemarks on several big spruces. The axemarks all faced toward the Meigs Post monument. After we had greeted the women, we all had faced the monument for a somber, almost prayerful moment. There stood a concrete highway right-of-way marker previously placed on the site by Rangers John Morrell and George Lamon in the early 1950s, crowned by a brass button naming identifying it the Meigs Post.

"Vinn, tell us what you feel now as you sit here."

"I feel I know these giant men who were here at the birth of our great country. These were men with "the bark on," who could go up the creek, over the mountain and down the other side with you. As I've followed their trail, I've realized they were extremely intelligent and hard as nails."

Vinn took off that big, black, ten-gallon Hoss Cartwright hat he always wore. "Dwight, you have ancestors from both sides of the Smokies. What are their assessments of Meigs Post?"

"Well, Vinn, my mother's family are Burchfields from Robbinsville. They are descendants of Chief Awahokee. The Cherokee word means keeper of the sacred water. They label this place foreboding. According to Horace Kephart, author of *Our Southern Highlands,* it is a dark mysterious fog-covered forest and a place full of superstition for the Indians—to be avoided at all cost. I can imagine Hawkins and Meigs trying to reassure the Cherokee who were with them."

Vinn leaned over to adjust his sitting position, putting his hand down on a large stone. The moss slid off the rock. Vinn looked up to see everyone staring astonished at him. "What you yahoos looking at? Ain't you ever seen a man rest before?"

"Look! Look, Vinn! Under your hand!" I said.

"Well, I'll be hornswaggled! You folks see what's on this rock? Somebody give me something to clean off this rock."

Phil produced a toothbrush from his never-ending supply bag. After much brushing, the letters MP appeared on the large rock.

"Anybody got any chalk?" Vinn asked.

"Yes, sir," retorted Phil. "What color?"

Meigs Post rock, Mount Collins

"Dad Jim it! White, if you got it."

The chalk revealed a message from long ago. "Look here, y'all! Return Meigs left this for us to find. He left it so long ago! One hundred and sixty-six years ago, in fact—MP meaning Meigs Post. This rock has endured the years of snow, freezing rain, wind, bears and humans of all varieties. And here an old ranger who took an oath to protect these lofty mountains finds the message."

Vinn covered the area with forest litter so as to deliberately obscure the old message, lessening the chance of future vandalism. I remained to further camouflage with limbs the rock we had found.

"Let's all meander down to the Mount Collins Shelter. We'll build a fire and set up camp," Vinn said. "I'll tell this story at the appropriate place—where it happened."

* * *

Leaving Dwight to his work, I went on with the group. As we walked, I was glad the duct tape covered my exposed area where the bear had eaten my rear pocket. I could hear the talk behind me concerning the patched hole in my pants, but no one was brave enough to ask the old bull of the woods to explain.

At the shelter, the coffee was hot and the fire was warm. I began the story. "Well, folks, these mountains hold many secrets.

"The first American surveyor was Benjamin Hawkins. I've seen pictures of him in the company of Washington and Jefferson at the time of Washington's resignation of his commission December 23, 1783. Hawkins was here when this place was crawling with lions, bears and Eastern Woods Bison. He was commissioned in 1792 by President Washington to survey the boundary line between the United States and the Cherokee Nation from Fort Southwest Point, the southern most point of the United States, based on the Treaty of Holston. The treaty was intended to establish the territory south

Hawkins at event when Washington resigned his Commission as Commander and
Chief of the Army, Annapolis, Maryland, December 23, 1783
Back row, second from right: Thomas Jefferson, Front: Benjamin Hawkins

Cropped photograph of a color painting by John Trumbull: "General Washington Resigning
His Commission to Congress, Annapolis, Maryland." From the Detroit Publishing Company,
1880-1920. American Memory collections. Library of Congress

---

of Ohio, which later became Tennessee and define boundaries
between the United States and the Cherokee Nation to curb the
illegal flow of whites squatting in Indian territory. The survey began
in April, 1797. The next surveyor was General Andrew Pickens
who replaced Hawkins. Pickens continued the line in 1797, from
Hawkins Post to the headwaters of Little Hungry Creek, northeast
of Hendersonville. The next surveyor was Colonel Thomas Butler
who surveyed the Butler Line in 1799, from Hawkins Post to the
junction of the Little River and the Oconee River near Tomassee,
South Carolina. The Butler Survey ended—remarkably—at or
near General Pickens' home. The whites had complained that the
Pickens Line gave too much land back to the Indians. The folks of
Hendersonville proclaimed, 'You put us in Indian Territory!'

"Now we come to Meigs' survey. In 1802 Return Jonathan
Meigs, the Indian agent for the country, corrected the Butler Line,
which the Indians complained took too much from them. His

line extended from Hawkins/Meigs Post to the headwaters of the Little River at Quillen Mountain just southeast of Brevard, North Carolina. This survey was not followed by another until Colonel William Davenport's 1821 State-line (Tennessee-North Carolina) survey along the crest of the Great Smokies from Davenport Gap to Deals Gap.

"During the ensuing century, there were numerous property disputes that led to the need for additional surveys. The Hawkins/Meigs Post was the pivotal point—beginning point or ending— of several of these surveys or disputes. Different folks claimed it was in different locations. Contributing to the confusion was the theft of the post. An individual was accused of moving the post, at the behest, if not the direct order, of Little River Lumber Company. One time—miraculously—it appeared nine miles away on top of Cold Water Knob on Miry Ridge. It got moved back to its rightful and present place on Mount Collins by a patriotic citizen—a competent surveyor. The Hawkins/Meigs

Meigs Post, Mount Collins

Post was lost, moved, lost again, returned and disputed—again and again.

"I believe that a contributing problem was that there were too many 'Little Rivers' and other 'Little' waterways in the equation. There is Little River that heads up here in the Smokies at Wilson Falls, Collins Gap and Miry Ridge near Buckeye Gap. There is a Little Creek northeast of Hendersonville, North Carolina. It is the end point of the Pickens survey in 1797. There is the Little River end point of the Butler survey on the Oconee River in South Carolina in 1799. There is a Little River at Brevard, North Carolina, the end point of the Meigs survey in 1802.

"The thing that is interesting is that here is a location—The Meigs Post—that, by all accounts, deserves more recognition than is visible today."

***

After everyone retired for the night, Vinn sat by the fire with me for a while.

"Dwight, I find I cannot stop following the trail of these men. When I have followed each of these surveys to their end points, then I will finally be able to let go. Will you come with me? I propose walking the Pickens Survey until it exits the park. Then we will drive the remainder of the Pickens Line, visiting different points along the line."

I had only to ponder the offer a moment. "Why sure! I'll go on the Pickens Survey Line, Vinn. For who knows! I may never pass this way again. Opportunity lost may never be regained."

The next morning, after reviewing our 7.5-minute quad maps, spread out on the school gymnasium floor, Vinn and I gathered in front of our small group of supporters. "Today, Dwight and I will continue the survey line S76°E to its terminus on Little Creek, Henderson County, North Carolina."

Julia and Vinn's three sons were not surprised to hear this from him. They reluctantly agreed, although each one had wished for his decision to be otherwise. They all put their arms around each other in silent family prayer.

R.J. and Chief Birdtown looked at each other, slightly smiled and discreetly nodded their heads.

Phil took a KJV Bible from his pack. I saw him open it to Psalms 121. I knew that one:

> *I will lift up my eyes unto the hills . . .*

With that, our supporters departed to shuttle my old 1956 Ford to the Luftee Visitor Center.

All knew that as we prepared to embark from Hawkins Post, we would be following a different leader of this survey. Pickens had been with Hawkins all the way from Fort Southwest Point to this spot. In July 1797 Hawkins had been recalled by President Adams to go arrest Governor Blount.

SECTION TWO

PICKENS LINE

General Andrew Pickens, the second in command, assumed responsibility for the remainder of the survey. Pickens had been with Hawkins all the way from Fort Southwest Point to Hawkins Post. In July, 1797, Hawkins had been recalled by President Adams to go arrest William Blount for treason. He had been governor of the territory south of Ohio and had later become a Senator. This information had been gathered by Vinn before the trip, from the Fort Southwest Point materials. Fort Southwest Point was the end of the road when one entered the "territory south of the Ohio River." It was no wonder it became a site loaded with historical documents. Roads were under the auspices of the United States Post Office. They were largely postal roads. He lived most of his life on his plantation called "Hopewell on the Keowee"—a short distance below where the Little River joins the Keowee River.

Painting of General Andrew Pickens
Photograph courtesy of South Caroliniana Library,
University of South Carolina, Columbia.

General Pickens was involved in many military conflicts and distinguished himself. He hosted many meetings at Hopewell between the Indians and the United States for the purpose of establishing treaties and cessations of lands—a term used when Native Americans ceded their territory to the United States.

Vinn was starting to pull ahead of me. He called back, "What was that they called him, lad?"

"The Cherokee called him Skyagunsta, The Border Wizard Owl." He was born September 13, 1739, in Paxton Township, Lancaster County, Pennsylvania. He died August 11, 1817, at Tomassee, his mountain home near the Oconee River, Pendleton District, South Carolina.

***

At Hawkins Post, not far from Collins Shelter, I set the staff compass in the ground, and Vinn led off into North Carolina. At the crest of the mountain, he descended down to Clingmans Dome Road.

Setting up on the lower edge of Dome Road, I watched as Vinn disappeared into the mountain ash, laurel, spruce and balsam. It was a laborious task, going through the nearly-impenetrable-to-jungle-like forest created by eighty-plus inches of high country annual rainfall. We made our way across a small hollow named Keg Drive Branch. The going got a little easier as the forest opened up. I could see that Vinn had the old spring back in his step. I pointed him up a laurel side slope that looked awful. Vinn dug in his heels, and like The Old Bull of the Woods that he is, climbed the mountain with vigor. Nearing an area of bluffs, I saw him stop, staring upslope. He motioned me forward. As I arrived, Vinn was flagging with surveyor tape the spot where he stood.

"Dwight, look up there! A cave! A big cave!" We climbed hand-over-hand. The cave was only visible within ten feet because laurel completely obscured it. The entrance was on a nearly-vertical bluff full of ivy. We were 0.8 miles from Hawkins Post—elevation 5200 feet.

As we pulled ourselves up to the entrance, Vinn suddenly was still. "Lad, there's a big copperhead sunning himself in front of my face. Lad, I dare not move a hair. Lad, you whup out that miniature umbrella you always carry. Stick it in front of my face slowly." I did as Vinn instructed, and I heard him quietly say, "Pop it, lad. Pop it *now*."

I popped it open and the snake attacked the umbrella. I could see it viciously hitting the fabric. Vinn "swarpped" at the snake with his wagon-tongue hiking stick, sending it down the side of the bluff . . . *toward me*. Luckily, it passed by, hitting the ground well below me. I had jumped sideways into a thicket of laurel.

"Vinn, you owe me a new miniature umbrella. This one is full of holes and snake venom." I could smell the damp, musty, sour odor of the copperhead poison.

Vinn called back, "I thought we got a memorandum from the chief naturalist that copperheads never go above 2000 feet!"

"Maybe that snake never got that memo! I think they follow their food source and stay to cool high places."

I climbed back up to the entrance where Vinn waited. There was an ice cold spring flowing from the inside of the cave. With flashlights in hand, we entered the darkness. There was much evidence that men had once worked this area. What they found, we could only guess. But lots of people had stayed there. They were Indian, not white. Flint chips were all about.

"Vinn, this must be Tsali's Cave." Tsali's Anglo-Saxon name was Charlie. He was a purebred Cherokee from the region and may have been part of Pickens' group. We knew that Charlie was listed as an old man when General Winfield Scott's cavalry came for him and his family during the 1838 removal. We knew he fled the troops and lived in a cave until sometime near 1840. Charlie would have been near 40 in 1797 when Pickens went through. He would have known this cave. At the cave entrance we could view the valley below for miles. Charlie would have been able to see anyone coming for him.

Vinn took in the area with a sweeping look. "Lad, let us leave Charlie's memory to the cave. Let us depart and continue our journey. This may be the dark, mysterious, fog-covered forest to be avoided at all cost that the Cherokee were referring to."

The spirits of Charlie and his strong sons may still dwell there since their execution by troops of General Winfield Scott in the early 1840s. After about four years of hiding out in the mountains, a bargain had been struck. If Charlie and his sons would give themselves up, the remaining Cherokee would continue to reside in this area of the mountains. Tsali and his sons were executed after they were taken into custody for the sake of their people.

We concealed the cave entrance, erasing any sign of our comings and goings.

We returned to the survey line, and Vinn led off. We proceeded on a side hill climb to elevation 5400 feet—1.1 miles from Hawkins Post. We sighted a compass line across the valley to Shot Beech Ridge. The compass determined that our destination would be a lone knob on that ridge. As we departed the 1.1 mile point, we descended down a very rough slope. Since we already knew our destination, I needed only to keep Vinn on line occasionally, using established voice and arm signal commands. At point 1.7 miles from Hawkins Post—elevation 4800 feet—we began our descent toward a valley. Vinn advanced a lot farther from me down the side hill at times. Whenever he was out of view, I would blow my police whistle, and he would hold up. No chance of his not hearing it. The noise emitted actually breaks the sound barrier.

Thusly, we proceeded down the ridge's left flank. We arrived at mile point 2.4—elevation 3600 feet—on Rocky Fork Creek. Vinn found telltale signs that Pickens had been there—the appearance of a very old camp. Remnants of old barrel rings reminded us of the enormous struggle it must have been to haul their food sources—including burlap bags of sweet potatoes and kegs of rum—through these mountains.

We descended the valley to Deep Creek at mile point 3.0— elevation 3320 feet—and decided this was a good area to camp for the night. It probably would have been the place Pickens camped.

It started out as a pleasant evening for both of us. Like the early surveyors, I baked sweet potatoes in the campfire—I am partial to them. *But we lack the rum, doggone it!*

Later that night Vinn and I had the pleasure of meeting one of the meanest bears in the park. Evidently this exact spot was his, and he made sure we knew that.

He entered the camp blowing, snorting, snapping and scratching. We grabbed our sticks and pointed them toward the bear, calling with guttural noises, "Hya, bear! Hya, bear! Get outta here." It didn't work.

I had noticed that Vinn was slow to anger. It took four hours of relentless harassment from the bear before his patience was expended. He turned over frying pans, rummaged in our bags and sniffed our lard. Finally, Vinn grabbed a burning limb and laid into the bear. All I could see were sparks and fur flying and occasionally heard a few words from Vinn. "This bear needs to know his limits!"

The next morning our camp looked like a war zone. I wondered if Pickens encountered the ancestor of this mean bear in 1797. I set the compass and pointed Vinn up the ridge, but he was already ahead of me. Suddenly, I heard a commotion in yon forest! In no time, out of yon forest came the mean bear followed by Vinn in hot pursuit. "This bear needs to know his limits!" The words rang out as the bear sped by, followed by Vinn swinging his wagon-tongue hiking stick in circles above his head.

I told Vinn to inform me when he was through playing with the bear. Vinn reassumed his position on the line, and we continued to the top of Shot Beech Ridge. We had to search a little to the east until we found the lone knob. Vinn found more sign there— old survey blazes on a tree. We were 3.4 miles from Hawkins Post—elevation 4200 feet.

We set our compass and sighted across a valley to Bee Tree Ridge. We paused to snack on cheese and apples. Vinn related the old legend of Shot Beech Tree. There had been a gold mine nearby where the Cherokee made bullets. The British had introduced them to firearms in the mid-1700s. The Cherokee melted old rifle balls out of gold and shot them for practice at the trees, then remelted them and formed them for reuse. They had shooting matches—which one could get closer to the center of the beech tree? An early trader named Borus Felmet would try to get the Indians to show him the mine. One Indian finally brought him to an old beech tree that was shot many times. Felmet was well advised that this tree was as close as he could get without being shot. The Indian left Felmet at the tree and was gone three hours. When he returned he had fresh chunks of gold to trade.

Leaving Shot Beech Ridge, we descended to Cherry Creek. From there we climbed to Bee Tree Ridge, elevation 4280 feet, 4.7 miles from Hawkins Post. The bee tree, most likely a black gum or tupelo gum, has a self-hollowing mechanism, making for excellent hives for wild bees. Bear hunters typically treed bears in black gum where they had climbed to eat the honey as well as the sweet droopy fruit of the tree. These trees grow in greater abundance when there are forest fires or clear cuts, the latter more common in North Carolina than Tennessee.

From Bee Tree Ridge we set the compass and shot across a valley to Long Drive Gap. When we arrived at Long Drive Gap— elevation 4800 feet, 5.8 miles from Hawkins Post—we located a bearing tree there that directed us toward Tuskee Gap.

Bearing tree, Long Drive Gap

We sidehilled to Tuskee Gap where we found a well-marked bearing tree, and we camped for the night. After supper we sat on the hillside looking down into the Luftee Valley. We could hear the river far away and see the lights of cars on the Blue Ridge Parkway, a national park area legislated to provide a "motor recreational roadway" between Shenandoah National Park and the Great Smokies. Vinn said, "Lad, so few of those motorists on that roadway are aware of the scenic rich history in the forests that they are passing so quickly." Vinn and I were so tired, we drifted off to sleep on the hillside. Out of my left ear late into the night, I could hear Vinn stirring slightly—then his voice, "Lad! Lad! We have company."

I sat up slowly and felt for my trusty hiking stick. *Nothing!* We had left all our stuff at Tuskee Gap, a good 300 feet away. "What is it, Vinn?"

"I don't know, lad. All I know, it's been sneaking toward us for a while."

I cupped my hands behind my ears and faced toward the intruder. I could hear the light crunch of leaves as the animal crept forward. Crunch, crunch, closer the animal came. "Vinn, I think it's an animal," I whispered. The thing came to within twenty feet of us. It was pitch dark. I did not know how long we had slept. I felt around the ground for a good-sized rock but could find none. *Where's a good rock when you need one?* Sweat popped out on my forehead as the animal came closer. I could hear the thing breathing. Then a low growl came from deep inside the animal. It seemed like an eternity passed by, then suddenly, "Grrr! Grrr! Bark! Bark!" A Bear Dog appeared in front of us and turned and ran down the hill.

Vinn was laughing loudly. "What was that, lad? Did that pup scare you?"

"You're dad-blamed right, Vinn! That dog had my attention."

We returned to our camp. Vinn secured his flashlight and went in search of the pup. "Lad, I think I know that Bear Dog. It's one of Chief Birdtown's dogs." We went back to our sleeping place on the hillside and, sure enough, found Chief Birdtown's dog there.

"Come on, pup, we'll get you some food." With that, the dog returned with us to our camp at Tuskee Gap and spent the night.

In my dreams I was enjoying the very best foot massage a fella could have. Then I awoke to see Chief Birdtown's Bear Dog licking the bottom of my feet. *Well, a fella has to get his foot massages where he can. That dog'll have a full day's work with Vinn's feet!*

Over coffee the next morning, we were favored with the company of Chief Birdtown as well as his foot massage dog. Birdtown said he had been training his dogs that night on Raven Fork, and the dogs chased a bear into the park.

As the three of us and the pup departed Tuskee Gap, Vinn commented, "The word 'drive' was repeated twice so far on the

Pickens Line—Keg Drive Branch and Long Drive Branch. That choice of term indicates a large group of men and horses—80 plus, in this case—were driving supplies against obstacles through the woods. This word is not used in names anywhere else in the area. This reassured us that we were on the line."

I could tell Vinn was feeling good because he was singing the old Christian songs as we walked. However, Birdtown's dog objected when I tried to sing also! "Aoooah! Aoooah!" he howled at the sky. We descended into the middle fork of Collins Creek, and the Chief showed us several marked trees before he headed home to Lands Creek. We very possibly would have missed them. The blazes were covered by laurel, rhododendron and their brush. Only the Chief knew they were there—from prior experience.

We crossed Collins Creek—elevation 3520 feet, 6.8 miles from Hawkins Post. We climbed an unnamed ridge which we christened Vinns Ridge #1, located 7.0 miles from Hawkins Post—elevation 4120 feet. Nearby, we found more old sign of the survey. We sighted across Luftee Valley to Vinns Ridge #2 and crossing Newton Branch, reached its 4260-foot elevation—7.6 miles from Hawkins Post.

We followed the compass reading, sidehilling down to the Newton Bald Trail located 7.9 miles from Hawkins Post—elevation 4200 feet. From there we went down Mount Clark to a point near Couch Creek—elevation 3400 feet, 9.0 miles from Hawkins Post. The sound of vehicles alerted us that we were approaching Newfound Gap Road.

After struggling through some very rough laurel, we came suddenly to a steep bank about 20 feet above the highway. This had been created when Newfound Gap was bulldozed in the 1920s. At the edge of the drop-off, Vinn struggled to get his balance and dropped his wagon-tongue hiking stick as he grabbed for a rhododendron bush. The eclectic Vinn hiking stick fell onto the edge of the highway, causing cars and motor homes to dodge and dart around it. The visitors did not see Vinn's boots dangling out of the rhododendron bush, which was hanging over the right lane of the highway, slowly descending to the road. I jumped down, muttering something like, "I'll save you, Vinn."

"Dwight, couldn't you think of something more eloquent to say?"

Waving frantically, I got the traffic slowed as the bush and Vinn settled onto the middle of Highway 441. Tourists had never seen a "Vinn in a bush" before, so it was their time to snap some memorable pictures to show the neighbors back home.

Some fellow from Georgia hooked a rope to the bush and dragged it out of the road. We sat along the river licking our wounds. Vinn had many more than I in that he was the one who rode the bush down the bluff. Several tourists joined us to ask if we were lost or stranded in the wilderness. Vinn said, "Lady, you don't know the half of it. We've been attacked by snakes, had to fight mean bears, were scared witless by a Bear Dog and fell through a bush into the middle of the road. Ma'am, you stay close to your car, and if any of these things attack you, jump in and drive off!"

Smokemont Campground was a short distance away. We hiked up the road to it and camped in the walk-in section. After supper, we strolled over to a ranger-naturalist program about birds in the Smokies. The snack shelter enticed me to visit often during the program. Vinn only wanted hot coffee. At the end of the program, the area ranger came over to shake Vinn's hand and converse with his old pal and me. He informed us that bears were frequenting the campground. He said the bears would come into the area near the walk-in sites, so to be sure, we stored our food securely.

I felt the urge to give an honorable mention to Vinn, the snake-bear-rhododendron fighter who sat in our presence. But alas, my moment had passed. The ranger said that one of the bears had ripped the trunk off a Karmann Ghia two nights earlier and jumped out of a tree onto a tent the previous night. "One more incident and we will transport the bear to a distant wilderness area where Karmann Ghias don't exist."

"Vinn," I said, "if that mean bear appears tomorrow, let me take care of him! Don't hog all the bear fighting." I spent the night shining my light out of the tent, anticipating the arrival of the mean bear.

The next morning, Vinn slept past nine a.m. I got up and made coffee. While sitting at the table at our camp, I saw a bear enter the edge of the campground. He was sniffing the air and looking around. I respected his rights to his space but kept aware of his location as I observed the human activity in the area.

I became aware that the park campground staff were preparing to clean the bathroom near our camp. I observed one employee come to the women's side of the bathroom. He knocked on the door and called out in a loud voice, "Cleaning Crew! Anybody in there? Cleaning Crew! Hello, anybody in there?"

Silence came from inside the room.

The employee turned the doorknob and pushed the door inward. He then placed a wooden sign between the doorknob and the doorframe. This left a space of about two feet with the door propped open. The sign read:

CLOSED FOR CLEANING.

I must not have had much to occupy my mind if this was my entertainment. I tried to picture this as a segment of *Days of Our Campground Lives* or *As the World Turns*.

With another hot cup of coffee, I started to return to my reverie, but what unfolded before me was better than any soap opera. I saw a lady approaching the bathroom door. After pausing to view the "Closed for Cleaning" sign, she grabbed the sign, opened the door and replaced the sign back in the door jam. The lady disappeared inside the bathroom. I could hear the cleaning crew gathering their equipment in the little room behind the bathroom. Buckets clanged together; brooms, mops and jugs of cleaning solution rattled around in the back.

Then, out of the corner of my eye, I saw the bear approaching with his nose up in the air, probably scenting some half-eaten lollypop in the women's bathroom trashcan. The bear rounded the corner of the building and approached the door.

"Vinn, Vinn, wake up! We have a situation here with the bear."

"I thought you were going to handle the bear problems today, Dwight."

"I was, Vinn, but this seems to be developing into a more serious scenario." As Vinn tugged on his pants, I explained what was happening. Vinn and I exited the tent. We saw the bear sniffing the air and poking his head in the direction of the half-open door.

I quickly dispatched the cleaning crew to run to the kiosk and get the duty ranger. The bear came closer to the door! Seeing the opening, he decided to enter. As his body went through, the door was pushed wider and the sign fell and hit him on the rear. In went the bear as the ranger arrived on the scene.

The door slowly closed, governed by a hydraulic arm on the top of the door. The duty ranger asked, "What's the problem?"

"There's a bear in the bathroom," replied Vinn.

"Well, we'll take care of that," replied the ranger.

"No, you don't understand. There is also a woman in the bathroom. A woman is in the bathroom too!"

"How did both of them get in there?"

"It's a long story, ranger."

The other two cleaning employees emerged from behind the building. "Hey, what's with our sign?"

"What if I told you guys you have a bear inside the women's room?" Vinn said.

For a moment, the two men discussed how they would remove the bear from the bathroom. Then the tallest of the two, whom I will call "Slim," said, "Hold it. We're not responsible for bears. We clean bathrooms. The rangers are responsible for bears."

"What if 1 told you guys there's also a woman in the bathroom?"

Slim thought for a moment and consulted with his partner. "Hold it. We're not responsible for women either. We'll clean up after the ranger gets the bear and the woman out of the bathroom."

The ranger, Vinn and I discussed the situation and came up with a plan. The bathroom was unusually quiet inside, considering the gravity of the situation. Vinn approached the corner of the building and bent over to the small vent that was probably six inches above the inside floor.

"Pardon me, ma'am. This is your friendly park ranger speaking. Don't be alarmed! Are you okay?"

"Yes! I'm okay, ranger."

"Ma'am, I don't want to alarm you, but we think there may be a bear in the bathroom with you. Can you ascertain the exact location of the bear?"

"A bear! A bear!" There came a slightly panicked voice from within. We were not exactly sure what happened, but evidently the lady did look into the next stall and observed a bear looking back. Our game plan went awry at this point, as all we could hear was blood-curdling screams coming from the bathroom. The ranger pulled his weapon and headed toward the door just as the door flew open, and out ran the hysterical lady.

"Ma'am! Ma'am! We'll need to get some information from you for our report—Who, What, Where, When and Why."

She was already a tenth of a mile away, her screams diminished only by the distance. We were not able to locate the lady, but later we noted that upon her rapid departure, she left several items behind in the bathroom, and we did notice a Karmann Ghia exiting the area. It never returned. "Well, Vinn. We saved the lady. You think we'll get a medal?"

"Dwight, you grab yon mop and climb up on this banister near the door. Slim, can you run fast?"

"Yup, I can!"

"Slim, when Dwight pushes against the door, you go up and turn the doorknob, then run like all get out 'cause the bear will be coming."

I pushed the door open, and out came the bear.

Slim was making good time, but Vinn had neglected to tell him the bear could run at 33mph. We observed Slim go out of sight with the bear closing in and passing him. The bathroom was flooding when Slim returned out of breath. Before departing, the bear had displayed his displeasure and fear by trashing every commode. "Well, Vinn, we saved everybody— just another day in the life of Vinn and Company."

* * *

We gathered our belongings and walked the highway toward Cherokee. Arriving at Vinns Bluff, as we had renamed the site of the rhododendron bush, we decided to split up—Vinn to finish the short distance to Raven Fork, and me to retrieve my vehicle and meet him at his destination. Raven Fork is the point where Pickens Line exits the national park. We had previously decided to drive the remainder of Pickens Line, avoiding the perils of civilization—dogs, cows, bulls and indignant landowners. We parted ways. I walked down the road to the Luftee Visitor Center where I retrieved my 1956 Ford—gray in color (the wording we use in a ranger report). Later in the day, I drove up the Raven Fork gravel road that abuts the park on one side and the Cherokee Nation on the other. I picked up Vinn where the line exits the park. We were 11.0 miles from Hawkins Post.

We commenced our driving phase of the Pickens Line. We were headed to the Ball Hoot Scar on the Blue Ridge Parkway. At the field behind the Job Corps Center, we paused as Vinn had a story to impart. "Lad, I recall we were having hunting activity late at night here. Wild boar were rooting up the meadow and jack lighters were frequenting the area. Illegal spot lights were being used to search for wildlife at night. We mounted an old stuffed boar on a piece of plywood and added lawnmower wheels. An electric winch and long rope were strung along the ground, providing motion. When a car came by shining lights, we pushed the button and the hog came out of the woods. The first night we were not used to the winch, and our hog ran across the field at a speed of 15 to 20mph and disappeared into the woods.

"'Man! Did you see how fast that pig ran?' one of the jack lighters commented. Right then, we accidentally hit rewind and our hog ran backwards in the field. 'Did you see that? That hog just run backwards.' 'I didn't see that,' replied the other. 'Here have another drink and maybe you can see a little better.'

"We returned several nights thereafter and got better and better at it. It's amazing when you cast a broad net, who gets caught therein." Vinn smiled and seemed to be reflecting on something amusing, but never further expanded on his thoughts.

My thoughts went back to the boar. The European wild boar were imported by the Vanderbilts in 1912 at Bob Bald in the Snowbird Mountains of Western North Carolina, near Robbinsville. The hogs were brought in for a 4,000-acre, private hunting reserve—sport for the guests. The hump of the boar is as high as three feet above the ground. The boar escaped and interbred with the domestic hogs and spread to other states. The Bear Dogs cannot kill them—the boars would disembowel them with their tusks. Only pit bulls can capture them. The dogs, bred for the very purpose of attacking boars, work in teams of three. Two dogs immobilize the boars by the ears and the third hamstrings a back leg.

We drove my old Ford for the rest of the Pickens Line. Our map showed the line climbed a ridge from Raven Fork to Mingo Creek at a point 12.2 miles from Hawkins Post. From there it climbed to near the Ball Hoot Scar—13.8 miles from the Hawkins Post. Ball-Hoot Scar got its name from loggers who would cut trees, limb them, point the heavy end downhill and yell, "Ball-Hoot!" The word comes from European loggers who arrived to help. "Beir-Hoot! Beir-Hoot!" they would call. It means get out of the way. A log is coming toward you.

From there we drove to a point near the Masonic Monument and walked in to see the beautiful stonework marker composed of rocks from many different countries. At the place where Pickens Line crossed the Blue Ridge, we were 14.1 miles from Hawkins Post. Vinn and I searched for bearing trees but found none. *Somebody probably Ball-Hooted the bearing trees down the mountain.*

We drove to Setzer Mountain and Campbell Creek. Vinn complained my car would not allow him to drink coffee. "Well Vinn, my ball joints are shot, so the wheels whimmy one way and then another."

"Above 45 mph, the whimmy is unbearable," Vinn said.

"It will help stir the cream."

I recalled that in 1943, a navy plane called a Bamboo Bomber got in trouble while flying over this very site. The controls froze, and the two pilots bailed out, leaving an enlisted man the task of riding the plane to the ground. Unfortunately, the man perished. Pieces of the wreckage still remained.

Plott Balsam Mountain near here was the highest point between Hawkins Post and Little Hungry Creek, the end of Pickens Line. The mountain towered 6200 feet. We spent the night in Hazelwood and the next day, ate breakfast at a local restaurant. We drove to a trailhead and hiked up to Three Tree Ridge. We could envision the three trees being part of the line, because Vinn recalled finding two other areas along the Hawkins Line that had three trees lined up. However, there was no sign at S76°E. We were 30.2 miles from Hawkins Post.

We left Three Trees and drove to Waynesville, where we knew that the line went right through White Sulphur Springs, but were unsure of its location. We went to the Waynesville Public Library and once inside—after looking around strategically—selected the most senior employee and asked for directions to the once-popular resort. It was clear that the silver-haired woman had been asked many times before. We were directed to turn left out of the parking lot, take Boyd Avenue to the bottom of the hill, go over the railroad tracks and turn left next to Blink Bonny Drive. "That's where the last shot of the Civil War was fired!" she whispered loudly as we were heading for the door.

The remains of the hotel were still visible. We could see the concrete pools where the white sulphur water would accumulate. "What did this water cure?" I asked Vinn.

"I have no idea. There were so many of these springs in the area. There weren't the pharmacologic remedies we have today for such things as high blood pressure and arthritis. People of means would seek out these spots, and after their ritualistic immersion with mint juleps in hand, would attest to improvement in almost any ailment."

"Vinn, as I was researching anything I could find about The Line in this area, I came across references to letters of William H. Thomas. History states that a letter was written in 1871 from William Holland Thomas, Chief of The Cherokee Indians, to John Platt, Waynesville, North Carolina. In the letter, Platt was requested to remove himself from Indian land he occupied at White Sulphur Springs, Waynesville, North Carolina, or a lawsuit before the United Sates Supreme Court would ensue.

"That's the same Thomas for whom the Thomas Legion was named, the guerilla group who fired those last shots in the Civil War." Vinn said. "And also the savior of the Eastern Cherokee Nation. He fought vigorously for Cherokee rights, property and citizenship. Thomas was seven-eighths white. Both he and Meigs were among the select white people that the Cherokee respected during that tumultuous time."

* * *

We drove to Bethel on 276 and en route, our highway was essentially superimposed for miles on the old Pickens Line. We parked at the intersection of 276 and 215 and got a cup of coffee at the intersection diner. I recalled an earlier day when I was in the area. "Vinn, one mile southwest of here is where I was involved in the recovery of General Wurtsmith's B-25 bomber in the mid-80s."

"Near Cold Mountain!" Vinn said.

"Who could forget? At Cold Mountain on the Canton side. It was one of the most beautiful places I had ever seen—an open, high-mountain meadow with wild azalea, purple rhododendron and blueberry patches. From Canton, we had taken 215 south to Bethel and continued on up 215 from this very intersection, then turned left onto Little East Fork Road. At the end of that road is Art Loeb Trail to Cold Mountain. It goes near the crash site."

Vinn spoke reverently, "Major General Paul B. Wurtsmith and all the people who died in that crash were American heroes."

* * *

From there we drove along the East Fork of the Pigeon to Big Hungry Creek. The creek headwaters are out near Frying Pan Gap, 42.2 miles from the Hawkins Post. It probably was named by Pickens after climbing Big Hungry and finding he had nothing to put in the frying pan.

Pickens Monument
Columbia, South Carolina

At the end of the day, we drove to Mills River and camped near the line. We were 49.7 miles from Hawkins Post.

The next day we drove to Little Hungry Creek, which is located northeast of Hendersonville, North Carolina. We had come to the end of the Pickens Survey, 65.0 miles from Hawkins Post.

For Vinn, this was a journey of 125 miles, and that would have been the same for Pickens. The difference was that Pickens had to walk the entire distance. Vinn walked from Fort Southwest Point to Hawkins Post and on to Ravens Fork, a distance of 60 miles.

We studied the 1884 C.C. Royce map and noted that the endpoint of the Hawkins/Pickens Line seemed to be at the headwaters of the Green River close to Ottanola Gap at the junction of the North Carolina counties, Rutherford, Henderson and Polk.

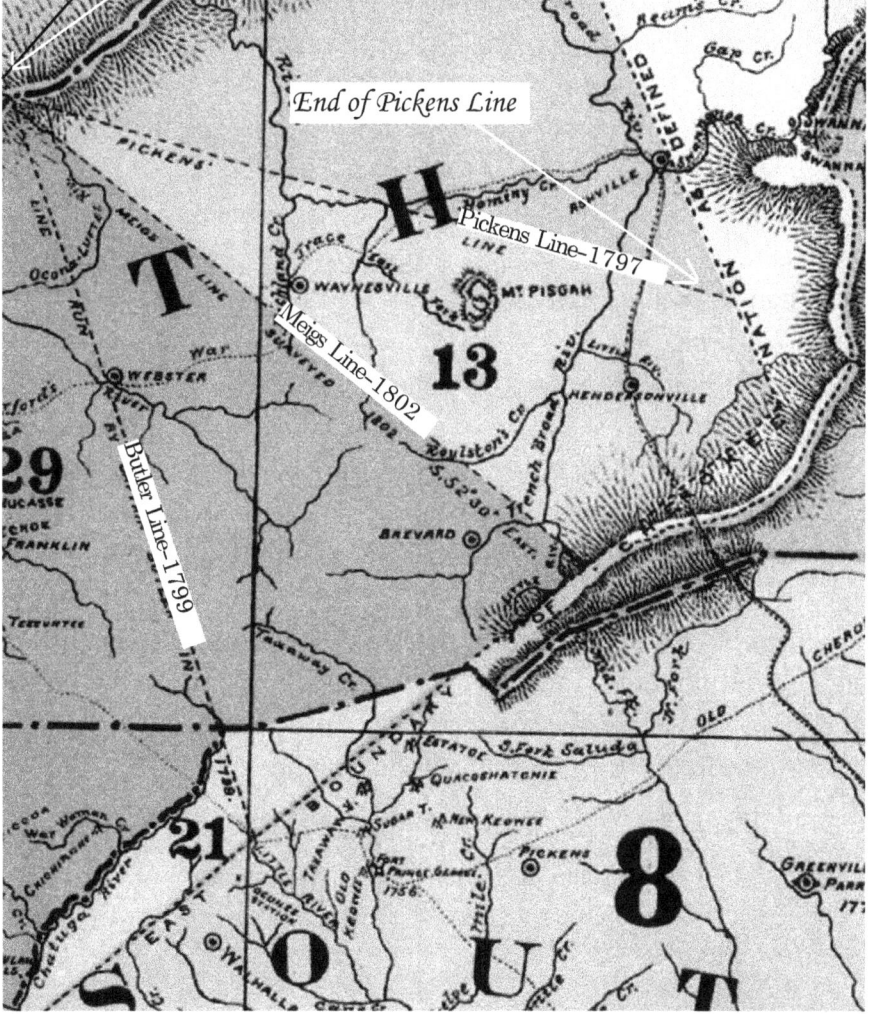

It appeared that the Pickens Survey left Hendersonville inside Indian territory in 1797. "Vinn, were Hendersonville and Brevard here in 1797?" I asked.

"Hendersonville and Henderson County were not established until the mid-1800s. William Mills settled in the area in the 1780s. The land occupied by the modern town, Mills River, and the land that would be the site of the present Hendersonville, were placed inside Indian territory by Pickens Line, making those early white settlers intruders and illegal. Their forcible removal was scheduled to occur two years thence. Brevard was established in 1851. The land the current town now occupies was also inside Indian territory, according to Pickens Line."

"Why was it necessary in 1799 to dissolve Pickens Line and resurvey as Butler Line?"

"It stands to reason that the settlers would have complained," Vinn said.

"I can hear them now. 'Git us back in white territory! Indians are movin' in!'"

"The Butler Line was the remedy. Instead of being forcibly removed, the new survey would legitimize their place of residence as being in colonial territory . . . but you know, lad, now we are obligated to walk that Butler Line so symbolically, Hendersonville and Brevard will be reunited with the settler territory, don't you think?"

"I agree, Vinn, and I'm available to go with you, but might we view the Butler Line from the car?"

"Yeah, lad, that will work, but let's take my truck. We'll walk the section in the park and drive the remainder."

"Where does the Butler Line begin in the park, sir?"

"Why, rat at the Hawkins Post, lad."

THE BUTLER LINE

I met Vinn at the Hawkins Post in the early fall of that same year for our planned trip on the Butler Line. The night before our departure we talked over coffee at the Tremont Ranger Station. Our family had gathered to see us off.

Vinn gave a brief history of the Butler Line. "The Butler Survey Line was named for Colonel Thomas Butler," he began. "Later documents refer to him as Lieutenant Colonel Thomas Butler. My thoughts are that Fort Butler at Murphy, North Carolina, is named after him. Colonel Butler is also mentioned in historical notes as a line officer at Tellico Blockhouse on the Little Tennessee River at Vonore, Tennessee. Colonel Butler was commissioned to survey from Hawkins Post at Mount Collins to the junction of the Little River with the Oconee River near the present Pendleton, South Carolina.

"The survey followed a heading of S27°E Grid N. Colonel Butler began the survey in the summer of 1799. The total distance from Hawkins Post to the end of the survey was 60 miles.

"Was that an extreme overcorrection at the behest of the white settlers?" I asked for the benefit of the group.

"Lad, I believe the survey was an attempt to reinstall what would become the towns of Hendersonville and Brevard back into settler territory."

* * *

Vinn followed me as we took his truck to Highway 19 at the Sherrill Cemetery and left it there to be picked up when we exited Smoky Park. Riding back up to Collins with Vinn in my old Ford was a pleasant experience. Vinn showed me that he had already drawn the Butler Line on the maps so we would not need to carry the cumbersome Jacobs staff. But we took it along in the vehicle as a backup.

Just as we entered the park at Cherokee, we encountered an extremely slow tourist in a Cadillac. Vinn said, "Pass him, lad, in that passing zone ahead."

"Vinn, I don't think we can!" I attempted to explain. "There's a prob . . ."

"Pass him, lad."

I peeked around the Cadillac and pushed gently on the gas.

"Lad, we'll never get around him this way. Floor it."

I floored it, and we went zooming around the Cadillac and kept accelerating.

"Lad, you can let off the gas now."

"That's what I was trying to tell you, Vinn. Lately the throttle hangs up wide open!" The old motor was doing things it had never done before. The valves were extremely loud. Smoke was coming through the floorboard as I was trying to lift the pedal with my big toe. A curve was coming up on us fast.

"Lad, shift into neutral and head for those young saplings." The motor whirred loudly as we crashed into the young poplars and sumac. Fearing my car would self-destruct, I jumped out, raised the hood and jiggled the throttle. The motor calmed down to an idle. I then saw the old folks pass in the Cadillac, looking at us with fear in their eyes. We got the old Ford back on the road and soon came upon the Cadillac.

I could see the lady conversing with her husband. "Dear, it's them outlaws again."

"Get off the road before they run over us."

Vinn gave a pleasant wave, acting as if we were on a Sunday afternoon drive.

We left the old Ford at Collins Gap and proceeded to Hawkins Post. We climbed on azimuth S27°E. Our first landmark was the Left Fork of Deep Creek, 1.3 miles from Hawkins Post—elevation 4400 feet. Our next landmark was Bearpen Ridge, 2.0 miles from Hawkins Post—elevation 5040 feet. We were on the top of a knoll on a feeder ridge just below the Noland Divide Trail. For the next several miles we followed the Noland Divide Trail and periodically crossed ridges over to the Butler Line, then returned to the Noland Divide Trail, which was more traversable. It kept us from going down into ditches and climbing back out and going down into hollows and climbing back out. We would sight a knob, go over and look and once again return to the Noland Divide Trail. Vinn located what appeared to be a bearing tree that had quite old markings. We had found that plain chalk would bring out the hidden messages.

From there we followed the line on the map to Burnt Spruce Ridge, 3.1 miles from Hawkins Post—elevation 4480 feet. There were indications of Butler Line there also, in the form of healed-over axemarks on old oak trees. We were now on top of a knob near another loop of the Noland Divide Trail. I could tell the present adventures had given Vinn back his energy and had put a spring back into his step. Given his ordeal near Wilson Falls, it was good to see the old ranger walking his way to glory—singing the old Christian songs—as we proceeded.

While sitting atop the high Smoky Mountain knob, Vinn told me of the roots of his Christian beliefs. "Lad, I have been on a lot of searches in these old mountains from just about the beginning of the park. I have searched for and found many children who strayed from the protection of their fathers and mothers. The joy in the eyes of one such child was indescribable when I found him days later, cold, hungry, alone and not knowing which way to turn. Lad, I too was lost in the wilderness of daily life. My sweet Jesus found me and offered me the redemption of his saving grace. Now he is my rock. I am no longer lost."

Vinn and I had a prayer right there on that mountain, and it seemed appropriate because, in the Old Testament, Moses and others met the Lord on mountaintops.

***

We left Burnt Spruce Ridge, descended to Pole Road Creek and camped there—elevation 3000 feet. We were 3.9 miles from Hawkins Post and had just crossed a park service trail. The evening was alive with night birds of all types.

The next day we climbed to a knoll above Deep Creek—at elevation 3200 feet—5.1 miles from Hawkins Post. We came to a huge bluff and overhanging cliff area. The cliff had formed a small cave. It appeared many different people had used this shelter area, possibly even during the Cherokee removal. We dropped down the mountain to Nicknest at Deep Creek and found no sign of Butler there. We climbed to Sunkota Ridge— elevation 3400 feet—6.8 miles from Hawkins Post. The line next took us to Indian Creek.

I can imagine the Cherokee living there in 1799 when Butler came through their homesites. *How did they react? Did they protest or challenge the surveyors? What would they say to the Cherokee who were helping cut the line? Did they conceal or hide or destroy the line later? I can only imagine. What man would stand by and watch his home taken?* The line was effectively moving their homes into settler land. We climbed to Thomas Ridge—elevation 3240 feet—8.5 miles from Hawkins Post.

We were at the park boundary and could see the old familiar paper and metal boundary signs. I had helped Vinn post a lot of boundary. The signs were used as targets by some folks. The bear and squirrels seemed to like to chew on them, also. The trees they were nailed to would spit them out of their bark in a few years and suck the nails inside their trunks. "Lad, God's creations such as these trees are no respecter of man's devices."

We dropped down into Cooper Creek and followed the line through the local community. We were 9.6 miles from Hawkins Post at elevation 1920 feet.

The local dogs had gotten word over the dog telegraph that intruders were approaching from a far land. They barked and growled at Vinn's wagon-tongue hiking stick. Vinn's sticks contained the old scent of hundreds of dogs from Kingston, Tennessee, to Hawkins Post, and from Hawkins Post to Little Hungry Creek, North Carolina. I couldn't blame them for being confused with all that dog smell. Even Chief Birdtown's dog had left scent on Vinn's stick. We arrived at our hike terminus at the Sherrill Cemetery at Highway 19 near Bryson City. We were 10.9 miles from Hawkins Post at elevation 1800 feet. Vinn's truck was a welcome sight as we put our sticks in the back. "We may have to 'coyote camp,'" I said. Vinn knew that meant lying on the ground and sleeping wherever the day ends.

Local dogs had given their scent to all of Vinn's tires, and one was intent on installing a final gift. Looking in Vinn's rearview mirror, I saw what I had never seen before—a young pup trying to scent-mark a tire while the tire was moving. "I've been here before, and you have to maintain a speed of above 30 mph to outrun him," Vinn said.

We drove to Bryson City to eat. *Lo and behold.* Vinn met some of the guys from Swain County Rescue Squad. We had supper with them, and Vinn reminisced about old rescues that we had done together. The fellows are real heroes; we never had to call them twice. They were having an overnight training session and invited us to spread our bedrolls at the rescue squad. *Boy, good to have friends when you are in need.*

After breakfast the next morning, we embarked once more on our journey. We followed Highway 19 up the Tuckaseegee River to the town of Whittier. We drove up the Thomas Valley Road to Thomas Cove. "Vinn, I wonder if this is named for Colonel Thomas of the Cherokee."

"Will Holland" Thomas, who was seven-eighths white and one-eighths Cherokee, was instrumental in acquiring lands, which became the present Eastern Cherokee Nation.

"Didn't he preserve and protect these lands for the Cherokee who were hiding out?"

"Yes, result of the 1838 renewal, Cherokee couldn't legally own land. Because he was 'white', Thomas bought large tracts and held it in trust for the Cherokee."

"Yes, Vinn, I remember reading that. By the 1838 agreement with General Winfield Scott, when Charlie gave himself up for execution, part of the solution was that the Cherokee hiding in the mountains could remain in Western North Carolina inside the reservation and not be subject to removal. Thomas held the property."

"What you didn't know, Dwight, was that Will Thomas was the adopted son of Drowning Bear or Yonaguska. Thomas was a hero to the Cherokee. A North Carolina historical marker honoring the

Chief is near his home close to Bryson City. It is located near the Kithua Mound, the first village built by the Cherokee in 1230. Originally the Cherokee called themselves the Children of Kithua. The white people are the ones who referred to them as Cherokee—keepers of the 'cherakee' or sacred fire. Before Desoto could enter a village, the shaman, who was the keeper of the sacred fire, would have to come out and perform a cleansing ritual for those wishing to enter. The sacred fire was in the center of the council houses."

We turned onto an unpaved road that took us to the very head of Thomas Cove where our maps showed that the Butler Line crosses. We were 13.5 miles from Hawkins Post—elevation 2000 feet.

Finding no visible sign of the survey, we departed. We continued upstream on Thomas Valley Road along the Tuckaseegee. At Nations Creek Road, we turned right, drove to High Point and found some survey telltale sign, old axe scars on dead trees. We drove back down Nations Creek, crossed the river and turned right on Highway 19. There we searched but never found the Log Cabin School referenced in the 1967 quad map. "Bet the four-lane took it," Vinn said.

We went back over the Tuckaseegee and climbed Barkers Creek Road. At the head of Barkers Creek Valley, we searched for survey blazes and found none. But the map showed—about 3.5 miles southwest of the line—there was a Blaze Gap, assuredly named because of the survey markings. We were 19.0 miles from Hawkins Post—elevation 3920 feet.

We returned to Highway 19 and drove east to Highway 441 at Dillsboro. A right turn on 441 sent us to our next point along the survey. Following Greens Creek, we turned right and went up the valley to its junction with Sugar Fork Creek. We found no sign there.

Back on 441, we drove to the community of Bucksnort, North Carolina, where the line crossed. After finding no ascertainable sign of the survey, we decided to stay nearby. We were 23.1 miles from Hawkins Post—elevation 2280 feet.

We ate lunch and pitched our camp for the night. Since this method of driving the Butler Line was less strenuous than walking, we decided to amble through the area and see the sights. A Wednesday service was about to begin at a local church so we slicked back our hair and grabbed some sweet shrub flowers and rubbed our underarms. We observed some gents chawing tobacco under the trees. "Good evening fellas. We are strangers sightseeing up your valley and would love to sit in on your service." We were welcomed wholeheartedly.

I don't know why Vinn asked the question, but he did. "Hey, fellas, has there ever been anything exciting happen here?"

"Why, mister, you ain't seen excitement until you've seen Red-Haired Mary come out of Skaloggadee Branch." The elderly gentleman seated in a rocker broke off a piece of homemade tobacco twist. "See that gravel road about one-half mile up the road? That's where she lives—plumb up at the head of it. Now you git up early and be on the front porch of the store up there . . . and you'll see something."

"You see, mister," chimed in a tall farmer with a straw hat and Camel overhauls, his attire complete with a white shirt for church. "Mary is always running a little late. She's the second grade teacher at the school."

"She got a new Ford from her family for graduating college recently," interjected a short, round gentleman with greasy hands who looked like a shade-tree mechanic. "Now that thar car will fly. It has a huge motor in it ,and Mary don't hold back on the throttle."

* * *

The next morning just before daybreak, Vinn and I were the first to take residence with the regulars on the porch step. "Vinn, you better park our transportation around back. I don't relish walking if Red-Haired Mary hits our vehicle."

It was a cold, moist morning. We could smell the ripening apples nearby. High up the valley we heard an engine roar to life. And then we could see a faint whiff of dust rising. Every once in a while we caught the glimpse of headlights. Then louder and louder the roar of the engine came closer. Down the curvy road came Red-Haired Mary. In each curve we could hear gravel flying off the road as the car fishtailed, first one way then another. Then I let out a gasp as three Bantee chickens wandered out into the gravel road. They were led by a smart-looking Bantee rooster who strutted his stuff for his hens. As he scratched in the road he suddenly stuck his head up at the sound of Red-Haired Mary rounding the bend and accelerating toward him. Although he made a good show of being courageous in front of his hens, he soon turned tail and ran down the road with his hens close behind. Red-Haired Mary—whether she saw any chickens or not, we can only guess—stomped the throttle and tore through the chicken family. We saw chickens twenty feet in the air. Mary never stopped at the paved highway but floored it and burned rubber for several yards down the road.

As Red-Haired Mary had come down the road in her high-pitched car, we had seen a slight smile on her face as she momentarily glanced in our direction.

"Vinn, did you see Mary glimpse me out of the corner of her eye and give me a look?"

"No, Dwight. It was obvious she was lookin' at me."

The old storekeeper sitting on the porch, said in a falsetto voice, "You're both wrong. She wasn't looking at neither one of you'ins; she always is a-lookin' at me every morning, and I've been sitting here every mornin' since she started driving that Mustang."

"Vinn, what about those chickens?"

"Dwight, those chickens live here. They know the risks of crossing the road."

With that, the Skaloggadee Valley was once more peaceful. I saw the rooster and his hens run quickly across the road with their heads down. The rest of our trip, that was one vision that stayed with me—the quick glimpse Red-Haired Mary gave me as she came to the paved road and then was gone, and the crimson red hair blowing out of the driver's window.

"Vinn, you need to make us a poem about Mary."

"Okay, Dwight, how 'bout this?"

> *Out of the valley came Mary with a roar.*
> *She had her foot all the way to the floor.*
> *Horses, cows and chickens alike,*
> *Took notice of Mary and her onward flight.*
> *Somewhere people set sipping their tea*
> *Not in Mary's hollow—It's chicken fricasse.*

Before leaving, Vinn asked the old head of the store if there was a spring nearby.

"Why shore, mister. Just up the road is Splattery Hollow. There's a cold spring there. The spring come right out of a bluff."

Vinn gathered his water jug and attempted to get a drink. The water comes out of the mountain off a vertical rock, and the area where it hits the ground is a lot of flat rocks. "Can't you get a drink, Vinn? You have to cock your head sideways to try." The spring was so splattery, he got more water on him than in him.

"Vinn, I believe the folks of Bucksnort enjoy a good laugh."

I hated to leave the valley of Skaloggadee, but we needed to try to amble on and forget The Girl with the Crimson Hair. We drove up 441 to Savannah Creek Road. The road took us to Pumpkintown, then to Long Branch, where we parked and walked to Windy Gap. We were able to find markings there defining the survey—two distinct healed-over blazes on line.

"Vinn, this is true to our experience. We usually find signs on the high points more often than otherwise."

"Right," Vinn said. "These are probably here because the survey shot from the previous high point, such as at Barkers Creek, to this high point. More evidence that the survey was not put on the ground between the high points of view. They shot from ridge top to ridge top. The jungle was too thick. It would be time-consuming to clear a cut line through rhododendron hells and other impenetrable vegetation. Furthermore, they would not want to lose identity of their high points because they might have to go back if they got lost." We were 24.9 miles from Hawkins Post—elevation 3480 feet.

We drove back to Dillsboro, then north to Sylva, where we picked up Highway 107 to Cullowhee. There we took the Tilly Creek Road to Cullowhee Gap. The survey line crosses right through the gap, so we could walk to it.

"I'm glad we reviewed this course," I said. Before traveling the Butler line, we had again spread out all twelve of our 7.5-minute quad maps in the gymnasium and connected the lines between Hawkins Post and the Butler survey end point at Little River on the Oconee in South Carolina. The end point we found to be just south of the 35th parallel of latitude. The survey lines had been drawn across the maps a few years earlier by Vinn as he dreamed of his obsession.

The 1799 survey had listed the end and beginning point as well as indicated compass bearing.

"Only grid north—or true north, using North Star readings—seemed to work with plotting the line because using magnetic north, which changes every year like the ocean tides, did not bring the line to the destination indicated in the history books. The journals of the surveyors discussed getting lost a lot, possibly because of the discrepancy in using these grid-versus-magnetic-compass readings, also possibly because of the proximity of a 'local attraction,' such as a rifle or another metal object which could disturb the accuracy of the compass needle."

We found only slight indicators of the survey—old notches, considerably smaller than the blazes. We were 27.0 miles from Hawkins Post—elevation 3800 feet. From Cullowhee Gap, we went to Thorpe Lake via Highway 107 and then on to Perry Gap. We parked and walked to the line where it passes near Wolf Knob, 29.5 miles from Hawkins Post—elevation 4800 feet. There are a lot of south-facing, high-mountain meadows there. They may have been produced by grazing, less likely by clear-cutting. In the vicinity, favoring the grazing theory, is the presence of Salt Rock Creek, the name indicating a salting place used to control the cattle so the animals would not leave the area for the whole summer.

The next check point was Yellow Mountain Church, which we reached by driving Flat Creek Road. We could not find sign of the Butler survey there. We took the Norton Creek Road to Higden Lake. The line crosses there, but human activity and nature appeared to have obliterated the markers. We were 35.7 miles from Hawkins Post—elevation 3800 feet.

We left Higdon Lake and took Greasy Camp Road toward Highway 64. Halfway down the road, we came to an unusual sight. A young groundhog had found a freshly discarded, number-ten green bean can. In his desire to get the last morsel in the bottom of the can he had succeeded in getting his head stuck. Vinn was laughing so hard, I thought he would bust. Bong! Bang! Bang! went the can as the groundhog wandered into one boulder after another. After a while, I could see that "save the groundhog" look on Vinn's face. "Vinn, how we gonna get that can off that groundhog's head?"

"Well, lad, I'll grab the groundhog, and you pull off the can."

The plan was working well until we implemented it. I tugged on the can, and the groundhog let out a sound, "Weeeee! Weeeee!" I kept pulling on the can, and the sound got louder.

That's probably why I didn't notice Mother Groundhog coming out of the kudzu. I felt an excruciating pain as Mother Groundhog bit me in the rear. Then I was screaming louder than the baby groundhog. The good Lord brought relief to everyone's suffering and allowed the can to come off. The mother let go my posterior, and I fled back to the truck.

A short drive brought us to Cowee Gap. We turned left past the golf course and came to Whiteside Mountain, then walked to the top of Whiteside where the line crosses and searched the rock formation for Butler's name. But, alas, there was too much mountain. The top is over two-and-one-half miles long, and there is a point on the ridge called Devil's Courthouse on the east end. Such a name implies it is so craggy and rough, only the devil would like it there. We had been told Desoto visited there looking for gold in the 1540s but failed to locate said treasure.

The nicest thing was the view. There was a sheer cliff, dropping about 800 vertical feet off on the southeast side.

Lots of tourists were visiting there. We were 39.2 miles from Hawkins Post—elevation 4880 feet.

We left Whiteside and drove to Highlands, North Carolina. A left turn on Bull Pen Road and a left on Cane Creek brought us to Whiteside Cove after we passed Granite City. The line comes through the middle of Whiteside Church. We located a tree that had some marks that were unreadable.

We drove south from Whiteside Church on Mill Creek Road. The road, being very rough, gave us a good taste of the mountains. We climbed through a high meadow then turned down a windy gravel road. The road dead-ended at Chattooga Cliffs, where the Chattooga River flows from the north in a southwesterly direction. A cabin sat there in a meadow, but no sign of any survey trees. The line passes right through the cabin.

This was some of the wildest country I had ever been through. We retraced our route to the Bullpen Road, drove southeast and crossed the Chattooga River. On the other side of Scotsman Creek, we came to the valley of the Bull Pen. The Butler Line crosses where the trailhead leaves for Ellicott Rock. We found no recognizable sign there.

We walked the Ellicott Rock Trail to Bad Creek and climbed to the top of Bad Mountain. We went slightly off trail and found Butler Line where it enters South Carolina at Bad Mountain. Plenty of sign was there—marking on line trees—and this reassured us our route was correct. We were 45.2 miles from Hawkins Post— elevation 2640 feet. We returned to the Bull Pen, drove east to Highway 107 near Mulkey Gap and then drove south on 107 to the Ellicot Wilderness Road, which took us to the Chattooga River again. No sign of our line there. The Butler Line crosses at what appears to be a canoe launch area.

We went back to Highway 107 and drove south. At Highway 28 we drove farther south to Walhalla, where we took Highway 271 past Pickett Post and Camp Oak and on to Tomassee, South Carolina. Butler Line crosses over the 1785 Eastern Boundary survey near there at the headwaters of Little River. We found Butler Line marked by several bearing trees pointing toward the Oconee River, 56.2 miles from Hawkins Post.

Before leaving Tomassee, we walked the 1785 Boundary Line for a distance. The line begins just below Tugaloo and has a bearing of 53° E Grid N. It goes northeast to a point three miles north of the Oconee Station—a fort where Pickens had been based for duty. The line then crosses near Cheohee and farther north. Then it crosses near Flat Rock Mountain in Pickens County, South Carolina, crosses near Line Runner Ridge and Indian Camp Mountain, crosses Pickens Mountain and then crosses the headwaters of the Little River, near Quillen Mountain in the Brevard area. The line then comes to the Little Creek in Henderson County, North Carolina. It ends at Tryon, North Carolina, at the Tryon Line from a totally separate survey. At this point the boundary line corners and heads in toward Asheville on a new survey.

Vinn and I noted there were a lot of Pickens and Little River names along these survey lines.

We drove east on Highway 271 to the junction of Highway 130 South, which we followed to Forest Road. We traveled along Forest Road to a point where the Little River joins the Oconee River at Lake Keowee, 60.0 miles from Hawkins Post. We were at the end of Butler Line.

We camped on the lake and took many a plunge into its waters. This part of our retracing of the old Indian surveys was over. But our planning for the upcoming trip following the Meigs' survey was just beginning.

SECTION FOUR

MEIGS LINE

Vinn and I sat before a campfire, the light reflecting off Keowee's tabletop waters on that calm night. Lacking reliable fishing poles, Vinn had used his wagon-tongue hiking stick tied to a short fishing line that was, in turn, tied to a Styrofoam cup for a cork. Fresh worms and spring lizards found along the Little River graced a bent U.S. Government paperclip hook, using the same technology that I used to fix my glasses. Being federal employees, we always had a pocketful of clips.

Having caught no fish, we hoped our luck would turn for the better on our next survey. After a time of quietude, Vinn announced, "We will need to return to Meigs Line for our last phase."

Colonel Return Jonathan Meigs
Photograph used by permission
from Meigs' family descendants

"Maybe we will indeed find whether the line was ever put on the ground between Meigs Mountain and Meigs Post," I said.

That seemed to start Vinn talking. "Return Jonathan Meigs was a man for his time. His journal was almost lost when, in the War of 1812, the White House was burned. We were cheated out of some of his story, but a portion of it was recovered, and we are grateful for it.

"Meigs was The Solver. He came along and resolved the conflict surrounding the two prior surveys here. He was the arbitrator who wanted to please everyone."

"Like Solomon, Vinn," I said, "he split the difference right down the middle to make everyone happy."

"But they weren't happy. They got upset even with his line. As you come through the present Sylva, you will be at the scene where Chief, The Bear, was upset because his village was now in white territory. Meigs would not relent. The only Western North Carolina historical marker that refers to any of these surveys is located right where The Bear stomped off in disgust and refused to bring sweet potatoes and rum for the survey party. And Meigs was the one The Bear had trusted.

"Meigs Post became the most important survey marker for the Great Smokies. It determined the boundaries in the initial deeding process of the park.

"Meigs Line became so dominant in the Park Service that 'The 1802 Final Survey' was the term used to collectively address the Hawkins Line, Pickens Line, Butler Line and Meigs and Freeman Lines. On the post were originally his name and those of the Indians with him. The post was six feet tall. It was illegally moved at the behest of the Little River Lumber Company in the 1930s. Six witness trees are still at the site."

"That answers our question, Vinn!" I had jumped up and was stoking the fire. "Why would the lumber company remove a post in 1930 that had been verified by the U.S. Forest Service survey in 1912? There are formal documents!"

"Perhaps the company did not know about the validating survey when they tried to maneuver the monument to their advantage." Vinn reflected his disdain. "John Morrell and George Lamon in the 1950s took a concrete highway marker, replaced it and put a brass button on top that had inscribed:

U.S.D.I. MEIGS POST

It too, was stolen."

"Vinn, you seem fairly knowledgeable about this subject."

"I agree I'm possibly obsessed. The information I have about the Meigs Line came from my mentor John Morrell. He was one of the finest men I have ever known. I helped him when one of his big tasks was to settle quitclaim deeds and other boundary disputes in the development of the Great Smokies National Park. In a quitclaim, two purported landowners would shake hands to dispense with the claim one of them had placed on a certain piece of property. The Smokies inherited hundreds of such deeds that needed resolution.

"I had to look into old prospect deeds in Elkmont Campground Area with references to hearings 200 years earlier based on Metes and Bounds property descriptions, which were common throughout the thirteen original colonies. Visual landmarks, events and distances were used to describe the boundaries of the property under consideration. An example: Beginning at the point where Elisha McCarter's sow swam river, thence across said river to the Winter John Apple Tree at the end of the corn row.

"Everybody 150 years ago knew exactly where that was, but it couldn't be defined today if one tried. But since it is a court document, it was accepted. John Morrell said, 'Tennessee law in the 1960s stated regarding the boundary lines that as long as there was a beginning and ending of the call—compass shot—to heck with the course and distance.' I paraphrase."

"The original landowners based their property descriptions on Meigs line, but had it ever been put on the ground?"

"I put the same question to my friend, John that you just asked me, Dwight. 'Do you think the line was ever put on the ground between Meigs Mountain and Meigs Post?' He said, 'It would be very tempting not to have done so . . .'

"My theory is that Meigs based the line on point-of-view to point-of-view, ridge-top to ridge-top, just like we discussed when we were on the Butler Line. The country is so rough. One clue is that in Meigs' journal, the focus was always at a mountain peak—Blanket Mountain, Bent Arm and Miegs Post—with no individual references between."

"Well, Vinn, that doesn't compare to a modern survey where the assumption is that there will be an unbroken line."

"This intrigued me. People shook hands, couldn't read, and here this line was never totally on the ground. For us, it is somewhat unsolved, but to the people of that day, it worked.

"There is another reason, in addition to the tough terrain, that the line was probably 'shot through the air.' The livelihood of each member of the survey team depended on being able to pack-train food and get the horses to the next meal. They were accustomed to amenities. So, when all is said and done, I suspect it was laid on the ground when there was easy terrain, but when the going got rough, probably 80-90% of the time, they took to the air, so to speak. It wasn't because Meigs wasn't thorough. He obviously was, as were the others."

I poured Vinn another cup of cowboy coffee. The last rays of the sun were gone.

"In the book *Strangers in High Places*," Vinn continued, "the author, Michael Frome, refers to a hike he took with John Morrell to Chimney Tops in the opening chapter. In the early 1930s, the U.S. Forest Service did a lot of research. The Great Smokies area was originally intended to be a U.S. Forest Service national forest but this project was abandoned, and it turned into a national park proposal. Present at the deliberations were a half-dozen people, including Knoxville's Colonel Chapman and representatives from Asheville. Some who were present, later participated in the Blue Ridge Parkway movement.

"Champion built a paper mill in Canton using equipment that had been intended to salvage spruce timber from Mount LeConte. The plan had been to construct a railroad across the state line and encircle Mount LeConte at a 4000-foot contour. It came within six months of happening. They logged up Sweat Heifer Creek almost to Newfound Gap. The park was able to acquire the LeConte land from Champion in that window of time. Therefore, the timber on Mount LeConte remains virgin to this day.

"Morrell acquired the Meigs information from the Archives at Washington, D.C., and from his father's notes. Norman Morrell was an attorney during the lumber company lawsuits of the 1930s. John Morrell gave me a copy, albeit partial, of the original journal of Return Jonathan Meigs in which he gave his account of the placing of the survey line."

"Tell me, Vinn," I asked, "from whence came the name Return Jonathan Meigs?"

"It's my understanding, Dwight, that in his youth, his father fell deeply in love with a special lady. He went to her to ply his offering of marriage. She immediately declined and did so again on each of the frequent occasions that followed. Heartbroken, after the last proposal, he turned to leave, with his head hung low. He heard a voice behind him. 'Return, Jonathan.' It was his beloved. 'I accept your offer.' The couple named their son Return Jonathan Meigs and Meigs named his son Return Jonathan Meigs, Jr."

The camp firelight flickered out.

***

The next day, we reclaimed my car at Collins Gap. After some R & R at our homes, Vinn's at Tremont Ranger Station and mine, at Tuckaleechee Cove, we gathered the logistics we would need to begin our retracing of Meigs Survey.

We planned to start at what is now called Meigs Mountain, just as Return Jonathan Meigs had done. This time we would be taking his journal. Once more, we assembled several 7.5-minute quads, this time depicting Meigs Line from Meigs Mountain, Tennessee, to Caesars Head State Park in South Carolina. Knowing the beginning point and the ending point, we drew a pencil line along the course, S52°30'E.

***

Vinn and I met for breakfast at his house. I looked over Vinn's shoulder while Sweet Julia ladled out the white gravy for the biggest cat-head biscuits I ever saw. He was studying Fort Southwest Point materials.

"The Meigs survey party gathered on July 25 in Fort Southwest Point, Kingstown, Tennessee, now Kingston, Tennessee," Vinn explained as he read. "In 1802, they left with a company of soldiers of close to 60 men as support, also twenty Cherokee axemen to serve as blazers of the line, and three Cherokee chiefs to ensure the participation of the Cherokee Nation.

"Vinn, let's check out the documents that show what they took with them."

"They had lots of rum, potatoes, bear meat, corn, bacon, oats, sugar, pots and pans, blankets and horseshoes for the twenty packhorses."

They had organized the caravan, gathered their equipment and people and went first to the base of Meigs Mountain. It had not yet been named for him, of course, nor had Blanket Mountain been named. Meigs was the one who placed the red blanket. The whole range was called the Great Iron Mountain.

At what became Meigs Mountain, our copy of the survey journal starts. It is the remains of Freeman's Journal that was recovered in Washington, D.C. Freeman was Meigs' surveyor. That is why some designations, in fact, refer to the survey product as the Meigs/Freeman Line. Meigs was the one commissioned for the project, and Freeman was hired as his surveyor.

"Vinn, look here now." I handed my 1884 Royce's map—entitled "The Cessations of the Cherokee Nations"—across the table. "It shows all three survey lines."

"You're right," Vinn said, studying the map.

In 1802, we had a known starting place. His line was to run from Meigs Mountain, which by today's landmarks would be about eight miles southwest of what was then White Oak Flats, currently Gatlinburg, Tennessee.

And we had a specified ending point, which was eight miles east of Brevard, North Carolina, on Quillen Mountain at the headwaters of Little River.

We moved our papers out of the way and consumed our last sit-down table meal for a while.

***

Vinn and I got our equipment together, and as reasonably as possible, we planned to duplicate the trip as Meigs probably went, starting where he started—at Meigs Mountain, eight miles northwest of Hawkins Post, later to be named Meigs Post.

After leaving Vinn's truck at the park line Boundary Tree—a tulip poplar tree in the center of the town of Cherokee, at the Dogwood Gift Shop—we drove in my car to Jakes Creek trailhead.

As we analyzed our maps, spread out on the hood of the truck, we felt that, to meet our goal of following the line as Meigs probably created it, the best route was to hike the Meigs Mountain park trail west one mile to Blanket Creek where we would proceed south cross-country and uphill. We would turn right at Bear Pen Gap. After encountering the line, we would go one mile back to the starting point. We arrived at our destination.

We shouldered our equipment—backpacks, Jacobs staff compass, food and a copy of Meigs survey journal transcribed by an archivist in Washington, D.C. Arriving at Blanket Creek, we walked upstream.

***

Having joined Hawkins Line, we reached Meigs' starting point and there found a promontory knoll—elevation 3900 feet. It is probably there that Meigs declared it to be Meigs Mountain. From that spot we shot across Bear Pen Gap to the next mountaintop, and established the point of view for our next checkpoint, a little to the left of the Blanket Mountain fire tower. We set our compass on S76°E from the blazed survey line and pointed directly to Blanket Mountain—elevation 4609 feet. We followed the line into the headwaters of Marks Creek and climbed out of the valley to Blanket Mountain.

"Vinn, when you came here on the Hawkins Line were you able to tell where the blanket had been stretched across?"

"I had to do a little searching. I recognized the size of a bluff on the line—a rock outcropping. It had to have been able to be seen from a backward look from Meigs Post. Looking forward, I could see Bent Arm, about three miles away between Blanket Mountain and Meigs Post."

We took time to visit R.J. Meigs Spring on Blanket Mountain, just southeast of the fire tower. Sitting at the R.J. Meigs Spring, Vinn and I looked out over the Marks Cove Valley—a splendid view straight ahead. The spring was cold and refreshing. The fire tower staff most assuredly must have used it. Gallons of crystal-clear water flow continuously.

"Back then folks had a knack for finding drinking water," Vinn said. "Meigs' guides would have known how to find this spring, later named after him."

We loaded our packs and returned to Bluff Rock—our name for it—where we believed the blanket had been strung.

"Vinn, look over there. Do you see that treeless green area in stark contrast to the dark forest around it? Don't you think that's the point of view Hawkins would have seen in 1797 and Meigs in 1802?"

"I believe the remnants of the meadow that exists there now would have been there—but in 1797 and 1802 much larger, given the fact that the buffalo have all been killed and the cattle no longer graze there since the park was established. The animals had kept it picked off and green. Now it has turned into a bald, typical of meadows long grown over that I have seen."

"Let's try our Jacobs staff compass and see where it points. Didn't you do this when you were here on the first leg of this adventure? Come and look through the horsehairs!"

Vinn bent over the compass and seemed to be deep in thought. I wanted his agreement. "You're right, Dwight. Nowhere else are you looking at a meadow contrasting green on black."

"Let's stay here tonight. I'm sure that is what Meigs did."

We set up camp there and reviewed Meigs survey journal entry dated August 13, 1802.

> 13th Ascended the Great Iron Mountain, traveling over very bad ground. Thermometer 52° Encamped.

"It is remarkable that he commented on rain and hail in late August," I said.

"It was most likely a wind out of the northwest," Vinn explained. "You've heard the term 'a cold wind in August,' haven't you?"

I was still studying the journal. "His thermometer did say 52 degrees."

"Got to remember the elevation," Vinn said. "Ice comes from higher up—eight thousand feet or more—and turns from rain into hail above the elevation of the Smokies. Reminds me of one time I posted boundary, when blue snow began to fall in a horizontal fashion. I was a young ranger in the fall season."

We had a pleasant supper and then, each with some of Chief Birdtown's corn dodgers and a cup of hot black coffee in hand, we climbed the tower steps and viewed the Great Smokies.

The next morning after breakfast, we reviewed our objective for the day. We got out Meigs' survey journal for August 14.

> 14th Iron Mountain. Traveled on the mountain 3 miles to find the object that had been taken and returned to our camp. Rained very hard, afternoon and evening. Thermometer 54°. Men packed provisions on their backs, leaving the pack horses that could not be got up the mountain.
>
> Paid Pierce 1 dollar & one dollar to Ryan. I paid one dollar to Unaketa & one dollar to Uhalouskoe $4.00. The service after finding that the pack horses could not be got up the mountain shall be one dollar per day until the hard service in consequence of packing men's backs is passed, to commence the 14th instant.

"Vinn, I believe they forgot to enter in the log that they had purposely tied a red blanket onto the bluff. I can just imagine what they would have said: 'Proceeded up Miry Ridge and once we came to Bent Arm Ridge, as we looked back toward the red blanket, we could see Blanket Mountain, but the red blanket was missing. This was confusing because we had tied it securely—we thought. We traveled three miles back to look for the object.'"

Vinn laughed, "I wonder if a bear came and, because of the food on the blanket, carried it off. He could have been attracted by the residue of the sweet potatoes and droplets of the rum—the official rations."

"I don't think they knew about the park rules to leave no food out," I said.

"Not to mention what the park officials would say about the restrictions on alcoholic beverages, Dwight."

"I believe Meigs must have dispatched his Cherokee guide who very efficiently tracked the blanket down and returned it shortly. They reattached it more securely on the rock and, when they returned to their camp, they could now view it in the distance. I suspect their new camp was at Bent Arm Ridge."

"It's a good thing that guide was not Chief Birdtown," Vinn said, "else there would have been bear meat to go with the sweet potatoes and rum that night." He studied the journal. "The men must have been tired after packing their provisions up to Bent Arm on their backs."

"Vinn," I said, "it appears to me that after finding that their pack horses could not attain the top of the mountain, they had to pack on people's backs their basic supplies. The packing commenced on the 14th. It most likely took all day to get the equipment to the top of the mountain. The record indicated they paid Pierce $1.00 to pack all day, and Ryan and Unaketa each $1.00, and Uhalouskoe $1.00 to total $4.00. What about that?"

"This could be the great Junaluska, Dwight, as he is referred to in the drama *Unto These Hills*. It would make sense that Meigs was writing phonetically and not the exact spelling. Was he of an age to pack supplies in 1802?"

"I did some research before our trip. He was born in 1779 and died in 1855. That would put him around age 21. He most assuredly would have been young enough to pack up to Blanket Mountain. Interesting and consistent is that he wasn't mentioned in the survey team until they got to Great Iron Mountain."

"He would not have been famous yet, but he was already known for supporting the white community." Vinn looked thoughtful. "I suspect Big Bear, listed in the Fort Southwest document regarding the caravan members and their supplies, is Drowning Bear or Yonaguska. He too was probably alive in 1802 in Johnstown. His adopted son, Will Thomas, was a colonel in the Civil War. Despite the similar-sounding names, I think Yonaguska and Junaluska were two different people. There is a marker for Yonaguska—Drowning Bear—located north of Bryson City, at the old village of Kituwah near where we saw the mound when we retraced the Butler Line. I believe it gives Yonaguska's birth year as 1750. He died in 1839, one year after the Trail of Tears. There is a historical marker in Robbinsville at his gravesite.

"Dwight, Junaluska and Yonaguska were friends despite their age difference, according to *Unto These Hills*. And it is interesting, but not surprising, that both of these fellas might have been on this survey."

***

We spent the rest of the night camped near the tower and, the next morning, reviewed the Meigs survey journal entry for August 15.

> 15th One dollar and a half paid to Lee.   1.50
> traveled six miles N.E. on the mountain & encamped;
> supposed to be near the object of view- taken from a
> mountain lying between the Iron Mountain & Blanket
> Mountain; the mountains being so much enveloped
> in the clouds this day as to prevent any distant view.

We left Blanket Mountain and headed for Bent Arm, 2.3 miles ahead on the line. There we commenced to locate the "back-sight" on Blanket Mountain, where the red blanket would have been. Mimicking the activities of Meigs and Company, we pretended a bear had run off with our invisible blanket. We determined we had lined up correctly with the "backwards" point of view, and thus, we were on the correct line.

We then shot from Bent Arm on a forward course of S76°E to Meigs Post and found it was 6 miles from Bent Arm, or a total of 8.3 miles from Blanket Mountain.

"Vinn, this country is so steep, we should have hobnails in the seat of our pants!"

"Lad, it still beats hiking with only one boot any day."

I stopped. "Vinn, I'm not so sure that Meigs walked it. Look again at what he said!" 'Journal entry: . . . traveled six miles N.E. on the mountain & encamped . . .' He said six miles northeast, but the survey line runs southeast! They are striking out northeast! They are avoiding this terrain. They must have known about the old Indian Gap Road."

"Guess what, lad, Indian Gap Road is exactly six miles northeast of where we are right now—this spot on Bent Arm. They must have camped at Indian Gap. I feel like we are solving a mystery!"

"We need to proceed post-haste to Indian Gap and camp," I said.

"No need to look for survey lines in this area because, just as I suspected, this line was never put on the ground between Bent Arm and Meigs Post."

"They probably selected their object of view," I said, "that old meadow, perhaps more distinct then than now, and then they must have joined back up with their horses at the bottom of Meigs Mountain on their way to Indian Gap. Vinn, do you think Hawkins did the same thing?"

"He referred to the point of view he saw from Bent Arm, on what later was named Mount Collins, but he never indicated how he got there; and now that you raise the question, I don't remember finding any markings between Bent Arm and Mount

Collins. It is easy to see how some land deeds could never be settled because no line could be found in the court records."

***

We proceeded down Bent Arm and arrived at Huskey Gap Trail. We walked up to Huskey Gap, crossed over to Newfound Gap Road and began plodding along uphill. In doing so we were paralleling the old Indian Gap Wagon Road where Meigs' team would have walked to bypass the Bent Arm-to-Mt. Collins line portion. Several tour buses passed, and the tourists waved at us, a rough looking pair. Vinn being the largest, I could avoid some scrutiny by standing behind him and his wagon-tongue hiking stick. A ranger's car that badly needed washing came into view. Blue lights flashed as the occupants spied the suspicious-looking hikers.

Out of the car, in a pristine uniform, clean tie and shirt, and with a smirk on his face, jumped . . . no . . . it wasn't . . . it was . . . Phil!

"Where are you going? You guys been hitchhiking? You know it's illegal, don't you?"

"No, we haven't," Vinn replied, "but we'd be lying if we told you the thought hadn't crossed our minds."

"You guys need a ride?"

We climbed in to save time and at Phil's behest, suppressing comments like, *Do a bear . . . in the woods?*

As we passed the lower tunnel, Vinn remarked, "Dwight, there's the old Indian Gap Wagon Road going up that valley to the right." That is where we had intended to resume Meigs route but the offer of a ride was too tempting. We would reconnect with Meigs' venture at Indian Gap.

***

Dirty on the outside but in better order than a lady's formal dinner table on the inside, the vehicle bore us toward Indian Gap.

Driving up the mountain, we neared Alum Cave trailhead. We settled in behind a small Toyota with a family of four inside—a father, mother and two little girls. All was well until suddenly the driver started throwing a fit. He slapped his head wildly, then the back of his shirt.

Phil pulled up and stopped right behind him because the Toyota abruptly stopped in the middle of the road. Phil radioed, "700, we may have an incident here. A man is whipping himself vigorously all about his body."

At that moment, the man bailed out of his car and did some sort of dance in the road. He proceeded to grab various portions of his clothes. All three of us approached him, ready to take evasive action.

"Ma'am, what's wrong with the gentleman?" Phil asked.

"Beeeee . . ." she replied rather calmly as she looked straight ahead and sipped her Diet Coke.

We were gradually able to put the story together. A bee had flown down his shirt at the back of his neck. He at first thought he had disabled the bee with a slap, but then it would sting him again and move to a different location. This lasted for a while until the man disrobed his upper torso completely and stomped every inch of his shirt in the middle of Newfound Gap Road.

As usual, Phil was prepared for anything. Out of his trusty kit he took several pennies along with his art-deco tape. "Here," he said. "You can put any penny on any bee sting and resolve the pain immediately."

As the children watched, wide-eyed, both groups returned to their cars and proceeded up the mountain.

***

We left the family with the bumble bee problem. Phil drove us up the mountain. It appeared Vinn was fixated on something near the top of "Ol' Smoky."

"Lads, look up there on the mountain. Looks like a storm brewing."

The closer we came to the top, the stronger the wind blew. Lightning flashed across our hood as cars descended the mountain slowly with lights on. Heavy rain and hail pounded our vehicle as limbs and leaves fell haphazardly in the road. Many campers and cars were congregated in pull-offs, awaiting a lull in the storm.

We arrived at the upper tunnel to find a white Dodge Dart blocking the inside of the structure. Vinn must have had a sixth sense about him. "You lads turn the blue lights on and turn the motor off. Then secure both ends of the tunnel," he said, hopping out of the patrol car. We could observe Vinn conversing with two persons in the four-door vehicle in the tunnel. Vinn shouted, "Phil, call an ambulance, stat!" Vinn disappeared into the back seat except for his protruding rear. His head emerged presently. "Phil, boil some water."

This is the first time I ever saw Phil speechless and confused. He looked my way. "Where am I going to find boiling water?" I pointed to a motor home parked near Phil. Phil disappeared into the motor home, emerging later with a pan full of hot water.

Vinn stuck his head out of the car. "Dwight, Phil, get up here quick." We arrived to see a lady in the throes of childbirth. Phil looked a little unsteady as he observed the situation. I grabbed the pan full of water as Phil turned pale as a ghost.

"Dwight, get my extra shirt from my pack." I retrieved an old ranger shirt and ran back to Vinn. He grabbed the shirt and held it open as the baby arrived with ease. Vinn tied the cord and folded the huge shirt around the little baby. The mother received the child and asked, "Girl or boy?"

"Ma'am, God has given you a healthy baby girl, in the midst of the wilderness, with a horrible storm raging."

"Where's my husband? We need to name her."

Vinn looked at me, "Where is the husband . . . and where is Phil?"

"Over there, sir. They are sitting, both pale as ghosts and in a full quiver. What you want me to do with Phil's water, sir?"

"Throw it on Phil and the new father. Maybe it'll wake them up." Vinn returned to the lady. "Ma'am your spouse is a little incapacitated."

"Ranger, what can we name her?"

"Well, Ma'am we are in a big storm and we're in a tunnel. Why not call her 'Stormy Tunnel?'"

"I love it! That's perfect!" The ambulance arrived and fetched the family off the mountain . We returned to our car wondering if we'd ever see Stormy someday at the Visitor Center in the Park.

***

Vinn and I stayed in the tunnel until the storm cleared, the wind and hail abated and Phil recovered. Eventually, traffic cleared, and Phil resumed transporting us to our destination, now seven miles distant. Turning right to Dome Road from Newfound Gap Road, we arrived at the old Indian route/wagon road/buffalo path. The road had been used for hundreds of years by humans and critters. We disembarked, retrieved our gear from Phil's vehicle, thanked him and waved goodbye.

"Vinn, it's time to break out the journal and see what's next." We studied the next entry:

> 16th. Thermometer 54°. After traveling three miles and trying various places found our point of view to correspond perfectly with two preceding objects, the first near 20 miles if the surface of the ground were measured over all the intermediate mountains to the position where a mound & post will be fixed tomorrow morning.

"Vinn," I said, "I think I know where the camp would have been. About two tenths of a mile from here there is a large flat gap. You remember that ice cold spring there on the Tennessee side?"

"As a matter of fact, lad, you took the temperature on at least one occasion."

"You're right! It's so ice cold, it nearly locks your jaw. Let's walk there and set up camp."

***

We arrived at the three-acre gap, a partially-open flat area that would have accommodated the entire Meigs caravan. It would have been 75% meadow in his time. We set up our tent, and, as the survey party would likely have done, we proceeded down to the spring to get water.

Having a thermometer in my pack with a range of minus 40° – 120° F, I placed it in the spring and watched it drop, finally setting at 41.7° F! I took it out, repeated the process and confirmed the findings. Having gathered our water in a large plastic container, we returned to our camp and proceeded west in the same direction the survey party went. We had a few hours of daylight remaining before we would return to our camp that night.

We walked 2.7 miles and arrived at Meigs Post on the Hawkins/Pickens line, which would have been the site of the survey team's referenced point of view and the place they planned to position a post the following day. Standing at the site, we surveyed the area and saw that an old meadow surely was here in 1802. There was mountain rye grass all about us, interspersed with spruce and fir trees. This was the first we encountered a grassy meadow since leaving the new campsite. Surely, then, this was the green area we had sighted through our staff compass when back on Bent Arm. As Meigs would have done, we moved about the area looking for a place to view the "back sight" and found an opening. In the direction of N76°W, we could clearly see Bent Arm, and Blanket Mountain farther in the distance. We strained our eyes to see what our maps indicated would be Triangle Mountain, approximately twenty miles back along the line, at elevation 2062 feet on the Chilhowee Mountain Range. The abundance of haze prevented us from seeing the most distant point of view that Meigs referenced in the 1802 report, a point he could apparently see perfectly. We could only see about nine miles back . . .

"Vinn, the difference in what we can see today indicates our world is changing drastically."

"I remember some of that change happening during my career," Vinn said. "Nowadays, I can only see half the distance on a 'clear day' compared to what I used to see."

We returned to our camp at Indian Gap. The next morning we reviewed the Meigs/Freeman Journal for August 17:

> 17th. Thermometer 51°. Ordered the pack horses to meet us on the line in the Indian country by going up the waters of the French Broad, Supposed to require about eight days. A spring at our camp being very cold was found by the Thermometer to sink the spirit to 48°.
>
> We marched 3 miles to a position on the G. Mountain, Erected a post of Spruce pine 15 Inches in diameter, Six feet high, Pointed at top, drawing a line from top to bottom to designate our course & marked on the north side U.S. 1802. R.J. Meigs A.W.D. T. Freeman, U.S.A. & on the south side C. N. U. and E., Cherokee Chiefs, erected a mound of Stones around the post of about 2 Tons of Stone, wh. with difficulty we collected having no Tools for digging. From this monument we commenced our line between the Cherokee and North Carolina and descended the mountain 45 chains and Encamped on its Side, Laurel being very thick.

"Lad, that must be the same spring all right that you measured last year. Meigs' thermometer may not have had as fine gradations as yours."

"What way do you think they sent the pack horses while they walked through rough country?" I asked.

"In my opinion, I think they would have sent them to continue on the Indian Gap Road to what is now Cherokee and beyond. They would have crossed the French Broad at what is now Little River Town near Brevard. I suspect they didn't originally plan to send the horses that way. They may have attempted to take the horses with them as they continued on the new line Meigs had been commissioned to do that represented a change in direction from the old S76°E to S52°30'E. The horses may have gone as far as that referenced campsite on the side of the mountain, explaining its current name—Keg Drive Branch."

"No doubt, they would have preferred to keep their rum with them in those rationed 50-gallon kegs, and the best way to carry them was by horseback," I said.

"No doubt, lad. I'll bet that when that caravan got all tangled in these laurel slicks, they were as confused as a termite in a yo-yo!"

"After the dense laurel encumbered their progress, they must have been quite reluctant to send the horses with the kegs on toward the French Broad. The term 'drive' in the name Keg Drive Branch, in keeping with the terminology from that time, referred to the transport of equipment or supplies by either wagon or horse. They had only horses referenced on this trip."

***

We broke camp and went back to the Meigs Post site where we had been the night before. We were eager to commence our journey along the new line.

I took Vinn's picture next to Meigs Post. The post was a right-of-way concrete highway marker. It had a brass button on which

Newer Meigs Post below hat

was inscribed "USDI," for the United States Department of the Interior; "Meigs Post"; and "NPS," for National Park Service.

Old wooden Meigs Post

"I would love to have seen that original spruce post," I said.

Vinn's voice was sad. "It may have been the one referred to in court records as having been moved at the behest of the Little River Lumber Company."

"I would love to have read first-hand all that information." I sat down and studied Meigs' report of the original inscriptions. "Listen to this, Vinn . . . we know the date was 1802. The letters after R.J. Meigs' name, AWD, would have stood for Army War Department. After T. Freeman's name the USS meant United States Surveyor." Vinn was looking over my shoulder. "On the South Side," I continued, "the Cherokee side of the line, the CN stood for Cherokee Nation. The letters U and E referred to Cherokee chiefs. I wonder if they were Unaluska and Cherokee Chief Elijah Ryan, the latter referenced in the Fort Southwest Point documents."

"Remember, lad, we found a dedication written on the back of the Freeman/Meigs Journal. They probably read it at the time they placed the post along with the two tons of stone. The whole monument must have taken quite awhile to assemble." Vinn set down his backpack and took out a crumpled piece of paper. "Let me read it now."

I stood up beside him. We removed our hats—Vinn's ten-gallon black hat and mine, an old National Park Service ball cap, and we paused in reverence. It was for a moment as though we were shaking hands with those early Americans—all of them. I detected a twinge of emotion in Vinn's voice.

> Erected by the order of Thomas Jefferson the Supreme Executive of the U. States. The man who in Eloquent Stile & firmly collected attitude, in the Face of the Universe announced the Independence of the United States of America.
>
> The undulations of his voice made Tyrants tremble, and revised the Hopes of the oppressed Humanity in distant Regions.
>
> August 17th 1802
> Cherokee Nation / Tennessee / North Carolina

We placed the staff on the ground and shot the next high prominent ridge on the azimuth of S52°30'E. At 7.6 miles, we saw a long ridge we identified as Thomas Ridge. The most prominent peak at the southern end was at elevation 4969 feet. The line passed just to the right. Our previously drawn line on our maps put the line at 4880 feet as it crossed the ridge.

"I see the same ridge as on the map," Vinn spoke from the compass.

We spent the night at Meigs Post, and in the morning we brought out the maps, spread them on a large flat rock and reviewed the lines we had drawn in the Townsend gym previously. We had a known beginning and a known ending for this leg of our trip. Meigs Post was at 6188 feet, and, on the azimuth reading of S52°30'E, the map showed the ending point to be on Quillen Mountain—elevation 3258 feet. This was eight miles east of Brevard, North Carolina.

"It is interesting to look at the first settlements recorded in the history books," I said, studying the map. "There were no white man's structures on the south side of the 'new' line. The first ones were built on the north side of Meigs Line. The first place to be built, according to the history books, was Morgan's Mill, followed by the Hackel Barclay House. Both were established in 1850. And Brevard homes were next."

Vinn said, "Think about it. At that time, this was the southern border of the United States!"

***

We descended the mountain on the line. We had gone only a few feet when we came across a pile of stones—mounded up, resembling a grave—that appeared to have been tossed there at some time past. They had wound up in a jagged line down the side of a hill. Possibly when the post was moved, I thought.

We continued down through large spruce trees and fallen timber. After 40 minutes we broke out onto Clingmans Dome

Road—elevation 5880 feet. We paused on the edge of the road to eat wild strawberries while we positioned our compass. We resighted Thomas Ridge as before. To be professional, we reversed our compass and shot back toward Meigs Post—N52°30'W— and it put us right on the site.

Moving southeast again, we followed our compass down what seemed like the steepest mountain in North Carolina. Traveling through dense jungle, laurel and fallen trees, we came to a "Y" where two streams joined in a hollow. We surmised this would have been where Meigs & Co. camped on the night of the 17th.

"That would mean that they went only a half-mile on the 17th," said Vinn.

"Let's check the journal. I think they went four miles on the 18th."

> 18th. Thermometer 63°. Sunrise. Continued the line down the mountain passing spurs of the mountain through very thick Laurel & Brier & encamped on the waters of the Tennessee.

"Vinn, I think that would be Deep Creek, 4.6 miles from the post." I said.

"I agree, lad. In 1806 the Cherokee would have called the Tennessee River what today we call the Little Tennessee. When the survey party used the term 'waters,' they were referring to headwaters."

***

We sidehilled following the line around ridge points and knobs— the spurs Meigs described. On one ridge top—near 4600 feet— we were surprised to hear humans talking. *Here we are, high on a mountain in rough terrain, and all of a sudden . . . it's women and men talking! The sound comes and goes.*

"Vinn, Vinn! I hear voices. It's women! It's men! Get out the map!"

Lo and behold, to our left was Fork Ridge Trail, within 1,000 feet of where we were. *Tourists!*

Then the sound of the talk trailed off. With regret, we continued on down the line, wondering who they were. Presently we came to the Left Fork of Keg Drive Branch. We saw an old path. I pointed out the sign of old traffic—up and down the stream.

"Some habitation has been here in the past," Vinn said. "Horace Kephart, author of *Southern Highlanders,* showed four homeplaces in this vicinity. He roamed the mountains of Western North Carolina, and his writings were very influential in establishing the Great Smokies Park."

We crested Fork Ridge near Deep Creek Gap, passing right across the park's Fork Ridge Trail. Vinn noticed four fresh footprints of two men (heavy, big-footed, walking in front) and two ladies (delicate, barely sinking into the earth, following).

Most likely made by the "words" we heard, I thought.

As we rested and ate granola bars, Vinn observed just down the trail, two groups of butterflies on the ground, circling on the path, and two more circles on the bank a ways above. "The groups on the lower side of the trail were where the ladies urinated," he explained. "The groups on the bank are where the two male companions followed suit."

"Where did you learn that, Vinn?"

"From Frank Oliver, the grandson of John Oliver of Cades Cove. He was one of the first rangers of the park and taught me so much."

"Why do the butterflies care?"

"The chemicals."

"Why were the ladies on flat ground and open trail and the men on the bank in the woods?"

"The ladies' area is usually positioned out of view but is typically right in the middle of the trail away from snakes and stinging nettles. The men would be off in the woods looking for a tree to step behind."

***

After the break we continued across Fork Ridge and descended to Deep Creek—elevation 2680 feet, 4.6 miles from Meigs Post. We were confident that it was the vicinity of the campsite for Meigs on the night of the eighteenth. But we had ample energy to continue. Leaving Deep Creek, we climbed and descended several ridges and crossed Indian Creek. Several hours later, we arrived at the crest of Thomas Ridge. The extremely flat high mountain ridge could well be the site of Meigs' camp on the night of the nineteenth.

> 19th. Thermometer 64°. Continued our course as yesterday over Spurs of mountain thro. exceeding thick Laurel & Briars & encamped on a ridge.

"This ridge was named after Colonel Will Thomas," Vinn said. "He was the person so instrumental in allowing the Cherokee Nation to remain in their ancestral homeland."

We pitched our tents, prepared our meal and settled down to a beautiful view of the town of Cherokee and Mingus Creek.

"This used to be called Quallatown." Vinn said.

"What was the story about Mingus Creek?" I asked.

"Abraham Enloe's farm was there. The Enloe Floyd tradition, as written in John Parrish's book *These Storied Mountains,* was that Nancy Hanks, Abraham Lincoln's mother, was purchased by Mr. Enloe to do work on this farm. Tom Lincoln came through on the way to Kentucky. Nancy Hanks was pregnant, possibly with Mr. Enloe's child. Tom Lincoln was paid to take her with him."

"That's it. Quite a legend!" I said.

"It was documented by the *Asheville Times* in 1885," Vinn said.

We reviewed the journal entry from the 20th.

> 20th. Thermometer 65°. Sunrise. Our provisions being nearly out, Sent our Interpreter & two Indians to the Bears Town to purchase provisions, delivering him Ten dollars & 10 cents. Continued our Survey on a ridge & on over a turn of it descending the side of a steep declivity over a stream 50 links wide, ascending and descending a high mountain to a fine stream of water & encamped on the West side of it. Killing two rattlesnakes on the route wh. Makes 5 killed since commencing the survey.

"They had to have an interpreter to speak true Cherokee language," I said. "Could that be Unaluska? We know he could speak English because, according to history, he communicated with the white people."

"That is possible. It was quite a point of honor. Very few Indians spoke English . . . if they spoke."

"They went to Bear Town," I said. "That's Sylva now, at Scotts Creek. The village was sometimes known as Johns Town also. The steep declivity would have been at Mount Noble. The lines on the topo map are pretty close together there."

"Fifty links is 33 feet," Vinn said. "The wide stream is probably the Oconaluftee River. A short distance upstream was the Boundary Tree, an early historical Cherokee monument, after which the gift shop where we parked my truck was named. That tree was a critical point for at least five surveys"

"Lad, I remember an older Indian who used to work at the Dogwood Shop, near the Boundary Tree. His name was Mason Walkingstick. He was in his later years of life when I was just beginning my work here in the Great Smokies. On one occasion he was pointing out the Boundary Tree to me, and he looked up into the sky and tears came to his eyes. 'Look, son,' he said, 'there's an eagle. You can tell that by the delta shape of his wings. A buzzard's wings are straight.' The skyward stare continued, as did the tears. His face was majestic and overwhelmingly sad. 'Used to see at least one every day! No more. That one first I've seen in years.' He looked at me, pointed to his chest and said, 'Son, a man should not take a picture of that eagle with a camera. He should only take it with this—his heart.' He walked down to The Oconaluftee River, steeply below the Dogwood Shop, where he sat down by the turbid waters, still with a sad expression. That was the last time I saw Chief Mason. I miss him."

Marker honoring poplar Boundary Tree, Cherokee, North Carolina

***

"They didn't camp at the Oconaluftee, they camped at a fine stream," I pointed out.

"That would be the Soco Creek. We won't be hiking to Soco Creek. We'll join up with the truck and drive there."

We exited the Park Boundary.

***

We drove south on 441 and turned up 19 and came to the Cherokee Baptist Church below Santa's Land. I noted from the map that the ridge between Oconaluftee and Soco was named Rattlesnake Mountain. We camped there, as Meigs had done more than a century earlier.

As we drank our cowboy coffee, turned our roastin'ears over the fire and ate corn dodgers, I marveled at the amount of snakes killed on this leg of Meigs' trip.

"And we can assume it's still richly infested," Vinn said. "Since the snakes live in large dens in this habitat, I can see how they saw so many."

"It was the dog days of August when they went through here," I reflected.

"You are so right, lad. Those Meigs surveyors must've had quite a time. During dog days, the snakes are blind. They have a film over their eyes while they molt. August is the worst of the three times a year when they shed. They strike randomly. They feel the heat of a potential prey but can't discriminate between human, squirrel or bird."

***

The next morning we were in dire need of a "cultural event." So we called Chief Birdtown. He recommended that we meet him for breakfast at Granny's Restaurant. It was located in the Soco Straight on Highway 19. This was where young Cherokee boys used to drag race their newly-acquired automobiles.

Birdtown had recruited many native Cherokee of differing positions in the tribal community to join us. We were inundated with stories, historical information and many interpretations of the survey. They gave both perspectives—how Indians viewed the line and how the white men viewed the line in the years that followed the survey. We shared the next entry of the journal with our old pals and new-found friends.

> 21st. Thermometer 63°. Sunrise. At 12 O'clock.
> 3 Indian men & 2 women fall in on our line with
> provisions; At Evening the Interpreter arrived with 3
> Indians bringing provisions. This day passed in site of
> several Indian Towns.

"Isn't that interesting," I said. "The women and the men were both burden bearers!"

Birdtown's chest rose in front of all of us, and his jet-black hair started to stiffen with great pride, as the other Indians looked on in awe. Taking his ancient pipe and drawing in deeply on the tobacco that filled the air with its pungent odor, he spoke. "See, our women are strong. Our women aren't weak. They bear the same burdens as our men."

Vinn, visibly moved by Birdtown's speech, said, "Chief, what do you think the food was that was brought in the provisions?"

"Mmm . . . bear meat . . . deer meat . . . but 200 years ago . . . it could have been you, white man!"

There was an outburst of laughter from the assembled Indians, one emitting a high-pitched whoop.

Vinn and I decided breakfast had come to a beautiful but abrupt conclusion. We exited the restaurant and reviewed the journal entry for the 22nd.

> 22nd. Thermometer 64°. Sunrise. Continued our line,
> varying a little to find the position of John's Town.
> and encamped near that place. were told that 8 or 10
> Indians were out looking for our line.

"Vinn, it looks like even though only two Indians went out along with the interpreter to get provisions on the 20th, there were a lot of Indians . . . a total of nine, presumably carrying provisions, that found Meigs on the line on the 21st, and now another eight or ten are looking for the line . . . as many as nineteen in all . . . bringing supplies."

"That ten dollars and ten cents must have been enticing," Vinn said.

We took old Highway 441 to Dillsboro. After trolling for family gifts in the quaint little shops, we went "upstream" on 23 to Sylva, where we found a road that followed Scott's Creek to the village of Addie, the town we believed to have once been Johns Town (or Bears Town). Meigs had arrived there on the 23rd. En route from Sylva to Addie, we located the historical marker commemorating the survey. It was about 200 yards off old 19 on a small tributary of Scotts Creek. We paused to read the inscription.

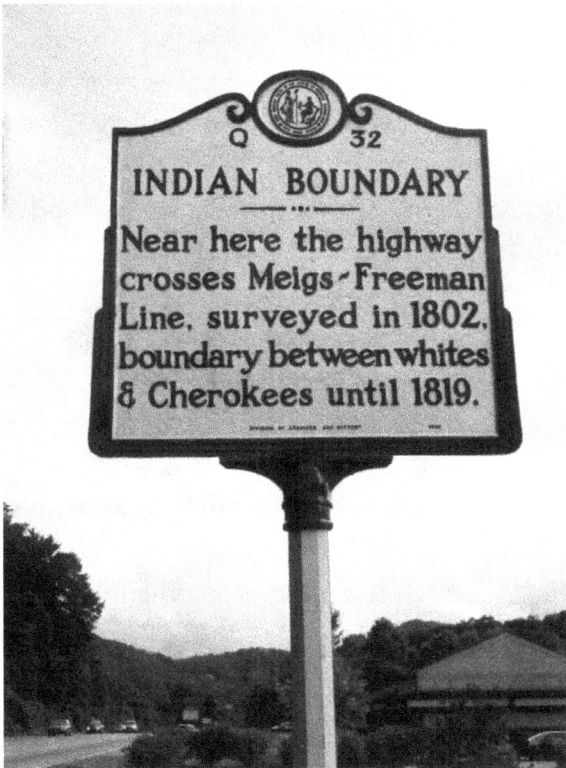

Meigs-Freeman Line Marker, Sylva, North Carolina

This was the longest-lasting line, and except for Big Bear, it was favorably received by most Indians.

We spent a lot of the day walking around Addie. We hiked to Ochre Hill Baptist Church. As Baptists are known to do, they were having a meal for a fundraiser and bake sale. There were lots of old stack cakes.

"Sweet Julia makes the best stack cakes," Vinn said, sampling as he spoke. "But these do compare."

"We make these with cookie dough and bleached apples," said the elderly lady behind the counter.

I dove into the banana pudding and downed it with fresh lemonade. They were organizing a quilting bee to provide quilts for victims of a recent natural disaster. After consuming all we could hold, we donated to the cause.

"Where do you go to church?" the pastor walked up and asked Vinn, while I conversed with several ladies nearby.

"First Baptist in Pigeon Forge," Vinn replied. They walked to a quiet area and appeared to chat awhile and have a prayer.

I asked the ladies about the names Ochre Hill and Scotts Creek. They sent us to look up some old heads on the Sylva courthouse lawn.

***

We located the elderly gentlemen playing dominoes. The consensus with the players was that Ochre Hill was a valuable site for the Cherokee, producing red ochre, a bright color used for pottery, war paint, dyeing ceremonial dress and many other things. They reported artifacts had been found near Addie that suggested an old Indian town, presumably Johns Town (Big Bear's town), had once been there.

The day ran into the evening. We were in need of a place to stay. We drove to Balsam Inn—on the Sylva side of Balsam Gap off the Blue Ridge Parkway. The appealing décor was little changed from the old railroad train days when the inn was a popular destination.

We sat on the veranda in the most comfortable rocking chairs we ever remember and enjoyed a view only a true poet could appreciate.

"Dwight, this old railroad depot here at Balsam, NC, sits at 3314 feet, and marks the highest elevation of any standard-gauge railroad in the Eastern United States. Up until a few years ago, folks would get off the train here and go right up the hill there to the Balsam Inn. Women would disembark with their big trunks of clothes to stay a spell.

"I like the design of this old station. It is just like the one in Sulligent, Alabama, where Dad, Mother, my brother and I would catch the midnight train when Dad worked in Tennessee. We would sometimes go there way after dark on a cold, frosty night to wait for the Midnight Special to Birmingham. We would be the only people there, and the place was never locked.

"There is more than likely a little coal-fired depot stove over there. Dad would build up a fire, and then we would go out onto the freight deck and push the handle on the semaphore pole in hopes to flag down the train coming from Memphis. I was four years old, cold and frightened, especially when the headlight of that oncoming steam engine would round the bend with the smokestack belching fire into the night as the train approached the station. Then the whistle would blow, and I would get behind Dad as if he could stop the train from running over us!"

I had never heard Vinn talk so much.

"A few years later," he went on, "we would move to the mountains where I hoped that railroads could not go. Delightfully, this little station at Balsam, as well as others, proved me wrong."

The menu of home-style food Vinn and I encountered was memorable.

"Dwight, the folks here at Balsam Mountain label this dinner, but I want to call it supper. That's what the evening meal would be for the mountain folks."

"This menu lists a salad of spinach and hot bacon," I said. "It sure sounds good to me right now and calls back memories of my grandmother's poke sallet, branch lettuce and creases."

"Are you talking about watercress?"

"No, Vinn. It's a cousin. My grandmother would find it in the spring, only in the fields where last year's corn crop stood. It sure made for good eating when she served it up with hot cornbread and buttermilk. She would parboil the creases in a kettle and then cook them in a skillet with bacon or fatback drippings. Sometimes she would add turnip sallet and crow's foot, also called toothwort, which the Cherokee called Indian mustard—ghuaree. Literally translated, that means 'place of the salad.'"

"Well, I think we encountered that salad place on Meigs Mountain. The white settlers referred to twin peaks as Currie He and Currie She. I understand that was their corruption of the Indian word, ghuaree."

I smiled. "I suspect that's because the Cherokee had a tradition of only packing 'she' plants because the male plant was bitter and the female plant was not bitter at all."

"How would they tell them apart?"

"Well, the men don't bear fruit in the plant world. You know that, Vinn. The men don't have flowers." I was enjoying this. "They only pollinate."

Vinn went back to the menu. "Anyway, that poke sallet sure enough sounds good! I bet these folks here would like to have that added to their menu."

I laughed. "Then they would sure enough call it supper!"

After enjoying a sumptuous fare, Vinn browsed in the library that boasted two thousand volumes. Meanwhile, I proudly put two pieces in the "community" jigsaw puzzle. I was joined by an eight-year-old, pig-tailed talkative little girl and a distinguished-looking elderly gentleman.

***

The next morning, after a wonderful breakfast, we reviewed the journal.

> 23rd. Thermometer 66°. Sunrise. Continued the line to John's Town where five Chiefs and a number of Indians were assembled who seem much concerned at the running of the line, having requested the Big Bear the principal Chief to go with us on the survey expressed much unwillingness to attend; but will give an answer tomorrow morning. The Indians discovered a friendly disposition bringing us provisions and fruits - more than we could take with us.

"I wonder what kind of fruit that was," I said.

Vinn remembered he had found a book in the library that was an old cookbook as might have been used in the 1800s. He fetched it. "Among the fruits used," he read, "were persimmons, blackberries, raspberries, huckleberries, muscadine grapes and mulberries."

"Don't forget fresh Paw Paw," I said, suddenly remembering my grandmother's teaching. She was of the Cherokee Clan in Yellow Creek, North Carolina, near Robbinsville. "It has the taste of banana and pineapple in a blender, not to mention it being an aphrodisiac—to be consumed at one's own risk. My granddad consumed too much on one occasion. He got so excited and rocked so fast, he rocked off the front porch and hurt his head. 'What happened to Pa Paw?' I had said. 'Well he ate too much Paw Paw.' Grandma Hettie had replied, 'The old man oughta know better!'"

***

We returned to Addie to rejoin the Meigs Line, and feeling quite content that we had clarified the fruit matter, we delved into the journal again, sitting at the train depot in church-pew-like uncomfortable benches.

> 24th. Thermometer 68°. Sunrise. Met the Bear & 4 other Chiefs to renew the conversation respecting wh. they were apprehensive would leave their Settlements on the Carolina side. They used every argument and [behest] for us to use our discretion; & run so as to leave them on the Cherokee side. They were told that it was not in our power, and that we were ordered to run the line from one given point to another given point; on our holding to this, the Big Bear refused to attend on the running of the line as he was requested to do. We told him that / in the / if event of their houses being left out we [would] make a representation of their situation to the Executive and did not doubt but their case would be favorably attended to by the Government.

"What a bureaucratic response!" Vinn threw his arms up and rolled his eyes back. "No wonder Bear stomped off!"

"This reads like the whole thing comes to a head on this day, August 24, 1802," I said. "The question was, 'Will my people, my cow, my house and my town be on the Cherokee side or on the white-people side?'"

"That's right, Dwight. Bear would have been the one to ask that very question. They called him the Big Bear so he had to be of imposing stature as well as a leader possessing empathy for his people. He was much loved and respected. That was understandable; he put his people first. Here he had led his people in providing massive amounts of fruit to placate these surveyors. It was his home, the town was in his name and it was the land of his ancestors—a spiritual place. After all that, the response had to be disheartening. Yet Meigs was a friend of the Indians. A lot of them believed that he was a fair man. He told them he would carry their concerns to Washington and plead their case to the president and told them to expect a favorable response."

"That remedy evidently did not satisfy Bear."

"Yes, they had wanted him to join the line at this point to continue with them on the survey, but he declined in a most dramatic fashion."

We read the last entry in the journal.

> 25th. A council of Chiefs met in the morning and agreed to send forward two of their men Lusena as Chief, the Big Bear having gone off in disgust.

***

As we prepared to continue our journey towards the South Carolina line, we lamented the loss of the pages that would have followed the journal entry of August 25, 1802. The remainder had been destroyed in the White House in the War of 1812.

"We may have Dolly Madison to thank for what we have of the R.J. Meigs journal," Vinn said. "She managed to save numerous documents as she and the president departed just ahead of the British. The journal wasn't entirely spared in the fire, but perhaps she went back at a later time and rescued further documents."

"Now we'll never know if Bear changed his mind and joined the survey team," I said. "Vinn, I've wondered in my heart, am I related to the Big Bear? I would be if Chief Awahokee and Big Bear's father or mother were kin. The chief kidnapped my ancestor Mary Burchfield in 1765 when she was age five. Her first child was Thomas, born in 1778. Thomas was my great-great-great-great-grandfather. Big Bear was the same age as Mary—both born in 1760. She would have known him and might even have been a playmate."

* * *

We studied our maps, spreading them out on the Addie depot bench between us. We looked ahead down the survey line and determined the next logical point for us to visit was Sugarloaf Creek, 23.6 miles from Meigs Post—3200 feet.

We went up Highway 23 and turned on Cashiers Branch Road. We found no line trees there. Finding no sign, we returned to Sylva and took Highway 107 and turned on Moses Creek Road— named for a tributary of Caney Fork Creek that flows into the Tuckaseegee River. We drove to a point near the Moses Creek East Fork and walked to the site where Meigs Line crosses. We again found no visible sign of the survey.

Leaving Moses Creek, we drove to Caney Fork Road, turned the truck east and continued on Caney Fork Road to Hawkins Knob. We parked near the Rich Mountain Church. We walked up to the knob—elevation 3729 feet.

"The word Hawkins couldn't be a coincidence here," Vinn said. "I don't think Benjamin Hawkins ever came here in his whole lifetime. Meigs must have named this mountain after him."

We returned to Hawkins Knob and soon found a marked tree with an axe blaze near Meigs Line, similar to axemarks found previously on the line and compatible with marks that are almost 200 years old.

We didn't need our compass to locate this tree. We knew the elevation on the line was 3729 feet on Old Bald Ridge, using our seven-and-a-half-minute quad-sheet topographical map. It is the largest scale map on which I am actually able to find my own location. Vegetation shading can be used as well as waterways, roads, buildings and other infrastructures.

Each elevation line on the map is 528 feet, so I could mark out how many chains I needed to walk up the hill to find the boundary. With one chain equaling 66 feet, my right foot would hit the ground 13 times. An individual's pace does not change no matter how tired one is, but steepness will affect it. A conversion can be made for the slope of the incline, right on site. Surface area on inclines versus on the flat was the subject of old debates. To avoid any confusion, Meigs was careful to indicate his distances were "as if the terrain were flat"—or as the crow flies.

Using an altimeter is a third potential adjunct I could walk up until the altimeter reads 3729 feet.

In Meigs day, even the change in temperature was used for location purposes.

"There's an old historical Judaculla Rock over in those cornfields," I said, pointing to the southwest. It sparked Vinn's interest, and he had to see it. We drove back down Caney Fork Road about three and one-half miles and after crossing a little one-car wooden bridge, walked across a cornfield to the site. You couldn't ask for a more peaceful place. We observed the many petroglyphs—carvings in a huge rock. Archeologists believe they were made 5000 years earlier.

"I heard the natives believe a legendary slant-eyed giant scratched the rock with his seven fingered hands as he leaped off a nearby mountain."

"Well, now, no foolin'," Vinn said. "With all due respect . . . it appears these carvings are pretty meticulous. I see on one symbol, a stone axe with sharp edges and also the symbol of the great-horned owl with its little tufts."

"I see a man walking stooped over, with long arms, and another man bending over a fire."

"This also has the appearance of a topographical map," Vinn said, "with river, roads and trails. I also see several nutting holes. They were used for cracking nuts with another rock, saving all the meat in the holes much akin to an egg separator."

"Well, I see a praying mantis here . . ."

"You do!" Vinn interrupted. "Well then I'm about ready to go back to the Cherokee legend."

***

Next we drove back to Sylva and headed for Rough Butt Bald. We entered the Parkway at Balsam Gap and parked near Bear Pen Gap. We walked the Rich Mountain Ridge to Charley Bald.

"That is how Tsali's name was spelled—Charley Tsali!" Vinn remarked, looking at the map. "The survey left behind more than blazes—names that have lasted as long as the axemarks!"

"I wonder if Camp Mountain ahead on the line is named thus because Meigs had a campsite there," I said.

We found old sign of Meigs Line on the bald: a blazed tree.

***

We referred to our map and headed for Pin Hook Gap on the Blue Ridge Parkway. At the Jackson/Transylvania line, we found old sign of the survey—a blaze at the ridge top. We drove to Cold Spring Gap, approximately one mile up. We filled our tin cups at the cold spring for which the site was named and reflected on the survey as we sat drinking.

"Vinn, there is a story that persists even today of Meigs' party being lost near here. The local tradition has it that Meigs got off the main survey route and wound up at Devil's Courthouse Ridge. He was reported to say that it was the worst possible terrain east of the Rockies. He said, 'Only the devil would inhabit a place like this.'"

"I suspect that is where it got its name."

"Yeah, Vinn, and it's also where the slant-eyed giant lived who jumped to the Judaculla Rock."

"Could it be possible that continuing down the line, Meigs shot from Pin Hook Gap forward . . . ?"

I finished his sentence for him. "Lookin' for the next ridge top!"

"Right, lookin' for the next big ridge." Vinn was deep in 1802 thought. "In order to get to the next ridge, they would have walked up to the top of Pisgah Ridge searching for what is now known as Shuck Ridge. I bet they turned down Devil's Courthouse Ridge by mistake—one right turn too early—and wound up two to three days lost.

"Let's drive over there."

"Well it is time to eat again. Chasing these slant-eyed monsters has made me hungry. Why not visit the Pisgah Inn at Wagon Road Gap for a festive fare? Meigs had to have seen Mount Pisgah. It dominates everything in this area."

While en route on the Parkway to the Inn, we stopped at milepost 422.4. There we hiked the moderate to strenuous one-half mile to the peak of Devil's Courthouse—altitude 5720 feet. It was on a trail that Meigs was not privileged to have. At the highest point was a panoramic view of the region.

"You can see how Meigs would be disoriented on this ridge," Vinn said. "It can be a 'land of perpetual twilight,' according to Dr. C. Hodge Mathis in his book, *Tall Tales From Old Smoky*. It gets dark earlier and stays dark longer when you are deep in the jungle forest of these mountains.

"He would have been using his compass, trying to connect with the line and the point of view from Pin Hook Gap and just got carried deeper into the forbidding wilderness."

We went on to the Inn. "It's my treat," I said. "I've heard the best is their chicken and their cobbler." We were right on time for the supper line and soon were seated looking south toward our destination. Bud and Danny, two rangers from Blue Ridge—brothers of the badge—arrived for supper and joined us at our table. We had shared many ranger incidents over the years.

"Do you remember," Danny said with a twinkle in his eye and his long-handle mustache twitching, "the vehicle I stopped near here with one taillight out?"

"Repeat it for us," Vinn said. "I can't be sure I remember it all." I was sure Vinn knew it by heart but wanted to hear it one more time.

"I turned on the blue light and pulled the vehicle over, and I informed the driver 'bout the faulty light. The man said, 'Ranger, are you sure the taillight is out?' So I escorted the driver back to the rear of the vehicle . . ."

"He had to be shown like everyone else these days," I said. "They all have an excuse."

"But he took one look, put his hands to his eyes and began to cry!" Danny said. "'Sir it's not really that bad of a violation.' I didn't know whether to offer a handkerchief . . ."

". . . or call for back up!" Vinn wanted to help tell it.

"The gentleman wiped his eyes on his shirt and wailed, 'Ranger it's not the taillight. Where's my trailer?' I had to turn away to keep from laughing out loud, but he appeared in such distress I contained myself. The poor fellow had departed from Mount Pisgah Campground thinking he had hooked up his trailer, but I didn't realize the magnitude of his distress until he said, 'Where is my dog?' and finally, '. . . and my wife!' It took me quite awhile to determine they had been in the trailer."

Vinn was laughing like he was hearing it the first time, and people at nearby tables were starting to stare. "I guess he should feel fortunate, since he could have been charged with transporting passengers in a moving trailer," Vinn said.

"Remember, Danny," Bud chimed in, "I sped from here to Pisgah looking for the trailer rolled down some bank, and there it sat in campsite A2 . . . the dog bounded out and nearly licked me to death. Then the wife came out and said 'I've been sittin' here for 35 minutes on the kitchen sink and the dog has about drove me crazy . . .'"

"I kept calling Bud," Danny said, "convinced there must've been a great wreck. Finally he came on, 'I found the trailer . . . muffle, muffle, snicker' . . . then he remembered he was on the radio and after a pause deepened to his ranger-voice level and said, 'Everything is okay. We solved the mystery.'"

* * *

We drove back west on the Blue Ridge Parkway and took 215 past Pin Hook Gap, looking for our next point of view. The line was about 200 yards to the southwest of our location on 215 as the road crossed the North Fork of the French Broad River— 2880 feet, 41.6 miles from Meigs Post. We parked and walked to Jakes Branch Gap to intersect the line and found no sign of the survey.

We walked back to 215 and proceeded southwest by truck to Daves Cove. We had an arduous hike from 2700 feet to Daves Rock, which is 3600 feet, 43.7 miles from Meigs Post. We had found sign of the survey—old age marks on one tree on the line—in Daves Cove on the way to the rock.

We returned to Shoal Creek and Balsam Grove communities. "The line comes right through the Balsam Grove schoolhouse," Vinn said.

I was more interested in trying to find out who Dave was and why he had a rock named after him. No one seemed to know. "It would seem an important place to have a rock named after you," I persisted as Vinn and I returned to the truck.

"Forget about Dave's rock; we have more work today," Vinn said.

We returned to our vehicle and drove due south on 215 where the North Fork of the French Broad River joins the West Fork. We turned northeast on Old 64 and came to the line at the point that Limekiln Branch crosses the highway. We found no sign of the survey.

We turned left on State Route 1331 and drove to Morgan Mill. This was the site of an early gristmill, established in 1850. It was a place of prominence for the arriving settlers—a place where corn could be turned into cornmeal. Stream banks by gristmills were the earliest places to gather in colonial days. They usually had a post office and were the site of political events and Sabbath gatherings. Such meeting areas served as the media for that day. The work of grinding the corn would be done during the week, and those waiting for the cornmeal would linger, smoke tobacco and talk.

"This was essentially the 'village green' of the community," Vinn said. "People brought ham and fresh vegetables and swapped them for cornmeal, which was the staple of the day. Wheat flour was less readily available."

"In fact," I added, "folks normally wouldn't use wheat flour unless the preacher was comin' to visit."

I commented that I had seen from my map that the river valley was settled early on. Morgan Mill was established even before Transylvania County.

"The river beds were the first areas settled in the whole country," Vinn said.

"Because of their accessibility and fertile bottom lands."

"Right, lad. The rich black soil would wash down from the high mountains and collect in the flat areas where it was enriched by occasional flooding."

***

We drove Highway 64 northeast to Route 1103 and came to Dunns Rock community.

"Who's Dunn?" I said. "And why did he have a rock named after him, too? Have we done passed it?"

Vinn rolled his eyes up, "And we're not even done yet."

We continued on 1103 and met highway 276 and turned southeast. At the community of Seeshore, we traveled on East Fork Road to the line and found no sign—50.8 miles from Meigs Post.

We drove back to Seeshore and turned right on 276. Approximately six miles down the road, we came to the community of Little River. At the Little River Chapel, we turned up the old Civilian Conservation Corps (CCC) Road and arrived at Quillen Mountain—elevation 3258 feet—and Caesars Head State Park at the South Carolina state line.

"Vinn, I want to know where Caesar's head is." We were sitting on a big rock looking back at the line, 56 miles to Meigs Post.

Vinn ignored me. "I don't want this quest to end."

I said, "I have a friend, Polly, who visited this park frequently as a child and as an adult, She relayed to me that it was a beautiful place to camp, walk the trails and wade the streams. She told me about the bears that live here."

Vinn's eyes twinkled, and his eyelids fluttered. "More bears? We've been fighting them all the way from Fort Southwest Point!" He changed the subject back to Meigs again. "This is presumed to be where Meigs' journey terminated according to the map."

ROUTE OF THE HORSES

"They had been without the horses for a good long time," I said. "That meant they were without their rum kegs. We could assume the horse party had already arrived here, at least a day before Meigs reached this point. I bet they had the kegs ready to greet their arrival."

"We could make that assumption, hoping they hadn't done any consumption."

My turn to roll my eyes up.

"We can't leave our adventure with Meigs' team lacking their kegs and sweet potatoes," Vinn said. "Let's in mind's eye make the trip with the horses."

"Good idea." I said. "The journal stated that they'd meet on the headwaters of the French Broad River . . . and Meigs said it would take eight days."

"Then it had to be a known course."

"The burned journal deprived us of the rest of their notes," I said.

"Then all we have left now is conjecture."

"I propose that they took the same route as the older historical party of Hernando Desoto in 1540. I'm convinced Desoto used existing Indian trails back then." They had hundreds of soldiers and their horses and were herding three hundred pigs they had brought to barter."

Vinn rubbed his chin and looked skyward like he was looking at a bird. There was no bird there. It was his favorite pose when he was pensive.

"It makes sense for Meigs' packhorse team to have done the same . . . to have known the number of days, they must have followed a trail of some type."

"That's why I put them on the same route. I'm convinced they would have used it."

"What was it?"

I studied the journal. "They left Indian Gap on August 17, 1802, going to the headwaters of the French Broad River. Their first camp would most likely have been sixteen miles on Indian Gap Road at Wears Fort—at the mouth of Walden's Creek where it joins Little Pigeon River."

"There would have been settlers and soldiers there at the fort," Vinn said. "The soldiers would have known the route east to the French Broad River."

"Yes, Vinn. They probably left on the 18th for Camp Two, and 16 miles a day would be about right for the horses. Sixteen more miles along the route would coincide with Zimmerman's Island—thirty two miles from Indian Gap."

"Is it an island?" Vinn asked.

"It was in Desoto's day and in Meigs' day. Zimmerman's Island is the Village of Chiaha in Desoto's journal. It was Indian property, albeit not Cherokee, until the white man came. The Dallas-culture Indians were Mound Builder Indians, not Woodland Indians. They were already there when the Cherokee Indians migrated into the area. Desoto arrived there June 5 and left June 28, 1540, spreading small pox in his wake. The isolated people had no resistance and were devastated by the disease with no idea what to do. They killed and ate skunks, hoping the pungent odor would counteract the deadly disease."

"What's there now?" Vinn asked.

"You'd have to drive to Dandridge, Tennessee, and row out to the middle of Douglas Lake and dive. The oblong island is now underwater, a result of the TVA project in the 40s."

"Where would their next likely camp be, lad?"

"Camp Three would have been on August 19, 1802. I bet it was at Newport on the French Broad—48 miles from Indian Gap. Newport sits astride the French Broad River that flows north, northwest from Asheville and enters the state of Tennessee near Del Rio. Like Kingsport, Newport received its name because it was declared a 'port of embarkation' for goods and equipment entering from North Carolina. For an early settlement, the official declaration of being an entry port was highly sought after. It meant the town became a center to ship and receive goods."

"And Newport remained a port well into the twentieth century," Vinn said. "Ranger Mark Hannah of Cataloochee District told me about hauling apples on a wagon from Cataloochee Valley to be sold and shipped out at Newport. It took him three days. He drove the apples by himself, and when he got there, they would take advantage of him—just a teenager—and underpay him, compared to what his dad would receive."

"The pack train would have then proceeded south up the French Broad River passing Hot Springs, North Carolina." I was expounding with confidence. "Camp Four, I believe, on August 20, would have been at the big bend of the French Broad River, opposite where the settlement of Stackhouse, North Carolina, was later established in the late 1800s—62 miles from Indian Gap.

"That site to this day offers itself well as a campsite. The delta created at the inside of the curve is very flat."

"Vinn, I believe on August 21, after benefiting from the good rest and enjoying a breakfast of fried trout, they would have pressed on to Camp Five near the new community of Asheville, North Carolina, at the junction of the French Broad and Swannanoa Rivers. That would be about 102 miles from Indian Gap. Desoto came right through Asheville, and I'm sure they did, too."

"One of the original four families who settled Asheville was Patton. Sweet Julia's mother was Ora Patton. The Pattons owned the land from what is now the Vance Monument down to the river, the same property now traversed by Patton Avenue."

"Well, now, Vinn, you sound as if you about own the town. Let me whoop a little on you, sir. Twelve hundred years prior, the Cherokee name of that territory meant 'where they race.' 'You completely left out my people, white man,' Chief Birdtown would say. The Indian name was 'untaki yasti—yi.' I suspect the pack train entertained the townspeople with horse and foot races."

"Are they gonna eat?"

"At least they had some trading fare with their rum kegs . . . less interest in the sweet potatoes."

"On the 22nd they would have headed out for Camp Six. I believe they would have camped where Mills River joins the French Broad River near what is now Mills River, North Carolina, 102 miles from Indian Gap.

"There's a great restaurant at Mills River. I used to go there with Parkway Headquarters staff—the annual gathering of the

Brothers of the Badge for superb prime rib. At that time it was called the Coachlight."

"One more camp before we meet Meigs," I said. "Camp Seven would be at the current town of Little River, North Carolina, on the French Broad River, 118 miles from Indian Gap—about 12 miles north of where we are sitting now."

"There they would have waited a day instead of coming up the mountain," Vinn said. "Meigs Line ended at Mount Quillen, where we are now, at the then and now South Carolina line, but then they would have connected up with the kegs."

"When Meigs got there, the Cherokee with him would have literally 'tracked' the pack train down and they would have known to ask locals if they'd seen the group."

* * *

We were on the state line of South Carolina within Caesars Head State Park. "I suggest we locate a good spot to camp for the night," I said. "We need a bit of closure for this endeavor that has consumed our interests of late."

"I agree, lad. Here on the top of Mount Quillen would be a good spot to dwell in the afterglow of all three of these survey lines. To add to that, I would like for you to fire up some more of that 'cowboy coffee'—just the way Birdtown would cook it—grounds and boiling water combined!"

I soon had a "rice crispies" campfire going. That means that it commenced to "snap," "crackle," and "pop."

Vinn commented about how good the hot tin cup felt to his calloused hands as he admired the view of twilight overtaking the South Carolina piedmont. He took a taste. "Dwight, this coffee is outstanding! How did you ever manage to find that vanilla flavor?"

"Vinn, a vanilla bug found it's way into one of the two tin cups. I didn't think you would mind. My grandmother showed me this bug a long time ago. It emits vanilla extract when touched. I saw it in the cup when I poured the coffee. That's the one in your hand now."

"Say, what?"

"Well, you did say to mix it up just like Birdtown would!"

"Yea, I did! Now that's right nice of you, lad!"

* * *

Around two a.m. the next morning, I heard Vinn running toward his backpack and shouting, "Get out of here! Bear! Detour!"

"Anything wrong, Vinn?" I called out just as he tripped and fell, attempting to grab his pack from the jaws of the departing bear, which gave no thought to delaying its departure. "You could not complete this adventure without just one more bear, now could you?"

"Lad, that blasted bear took the last two-pound bag of flour I was going to use for our Indian Fry Bread this morning! And, besides that, it was the state flower of Tennessee!"

"Vinn, you know darn well that the Tennessee state flower is an iris! What are you talking about?"

"No, lad, my brother phoned me last year after the Alabama football game and said that he heard that the state flower of Tennessee was either Martha White or White Lily—he was not sure which."

"Well, indeed, White Lily is a flour but not an f-l-o-w-e-r." I said.

"Just what I need," Vinn said, "The Pillsbury doughboy correcting me!"

After all of this misadventure, we decided to go back to the truck and drive to nearby Cleveland, South Carolina, for a good breakfast. This we found at The Mountain House Restaurant within 200 yards of Caesars Head park headquarters.

There were some South Carolina State Park Rangers having a breakfast meeting—some more very fine Brothers of the Badge.

They welcomed us at their table and were very patient in listening to us tell about the high places, deep hollows and wonderful people we had encountered.

***

As we drove home, the silence was overwhelming for a while and then Vinn spoke. "Dwight, we have completed the third journey of our four gentlemen who passed this way so very long ago. I feel that I have come to know them personally, especially R.J. Meigs. It saddens me, in a way, to bring this to an end. I recall an applicable quote from *Ulysses* written by Alfred, Lord Tennyson:

> *I am a part of all that I have met;*
> *Yet all experience is an arch wherethrough*
> *Gleams that untraveled world, whose margin fades*
> *For ever and for ever when I move.*
> *How dull it is to pause, to make an end,*
> *To rust unburnished, not to shine in use!*
> *As though to breathe were life.*

"Dwight, I feel it is an obligation for us not to fail to share with everyone, the people and events that you and I have relived during this venture. This was the southern boundary of the United States! These people, whether native Americans or immigrants, left their sweat and blood over these very same survey lines. We were actually tracking *them* because we found sign and clues that they left in these mountains so long ago.

"Yes, we have touched the very same places they did. We have had the privilege of 'shaking hands' with Americans who walked this way before us."

EPILOGUE

Ole Vinn's retirement dinner was finally arranged. It was held at the old Wonderland Club Hotel in Elkmont. It was a warm summer night. Most folks were sitting in the porch rockers until driven inside by the "no see-ums" gnats . . . all except Vinn, his family and me. He was taking in a sunset view of Meigs Mountain to the southwest.

"Folks," he said, "I now know some of the pain that was suffered by the man for whom that mountain was named. I must say, however, that before this night is over, I expect to be more frightened than Hawkins, Pickens, Butler and Meigs combined because, Sweet Julia, tonight I am going to attempt to sing for you, the boys and their families."

I alone have been privy to hear Vinn break loose in a song— as he led me through the forest as a young ranger. I was quite impressed. For Vinn to sing to us rangers and the family publicly doesn't happen except on very special occasions.

Julia and the family looked at each other and smiled. I went inside and shared this unexpected agenda with Chief Birdtown, R.J. and Phil.

Epilogue

R.J. whispered, "Well, by gosh!"

Chief simply shook his head and smiled, while Phil asked me if he should go to his pack and take out his pipe organ to accompany Vinn. I was glad Phil had learned to laugh at himself.

After dinner and all of the peer-group speeches, it was finally Vinn's turn. He went to the podium with the support of his wagon-tongue hiking stick, and keeping on his wide-brimmed black hat, he spoke, "I love this park—every ridge, every hollow, every critter and every buck ranger in it—although I could have done without the gnats and saw briars. It has been a very happy, satisfying and rewarding experience to be a part of this National Park Service organization and to have rubbed elbows with so many fine, sincere and dedicated people. May God bless you, each and all!"

He then looked at Julia, nodded at the pianist, blushed a bit, took his hat off and began to sing a World War I song written by Stoddard King and Zo Elliot:

> *There's a long, long trail a-winding*
> *Into the land of my dreams,*
> *Where the nightingales are singing*
> *And a white moon beams:*
>
> *There's a long, long night of waiting*
> *Until my dreams all come true;*
> *Till the day when I'll be going*
> *Down that long, long trail with you!*

There was not a dry eye in the house. There was silence—then applause. There were smiles. There were handshakes. There were hugs. Indeed, may God bless us all!

APPENDIX

Colonel Benjamin Hawkins was born in 1754 in Warren County, North Carolina. As a young man he attended Princeton University. He fought in the Revolutionary War and was a decorated officer. President Adams appointed Hawkins Superintendent of all Indians south of the Ohio River in 1797. Hawkins died in 1816 and was buried in Hawkinsville, Georgia.

General Andrew Pickens was born September 13, 1739, and died August 11, 1817. He fought in the Revolutionary War and was a decorated officer. General Pickens accompanied Benjamin Hawkins in the Smokies and when Hawkins returned to Knoxville, Pickens continued the Pickens Line from Hawkins Post to near Asheville, North Carolina, on the azimuth S76°E.

Lieutenant Colonel Thomas Butler surveyed the disputed boundary between the U.S. and Cherokee Nation in 1799. He was a commander at Fort Southwest Point. He died September 7, 1805.

Colonel Return Jonathan Meigs was born in Middleton, Connecticut, December 17, 1734 and died January 28, 1823. He was buried at the Garrison Cemetery in Dayton, Tennessee, in Rhea County. In 1801 he was appointed Cherokee Indian agent at Tellico Blockhouse. In 1802 Meigs and Thomas Freeman reran the Hawkins line from Meigs Mountain to Meigs Post at the present Mount Collins. From there he ran the Meigs Line 60 miles to Brevard, North Carolina. It took 10 weeks to run the survey through land that he referred to as "the roughest country east of the Rockies."

REFERENCES

Banker, Luke H. *Fort Southwest Point*. Kingston, Tennessee: Roane County Heritage Commission, 1984.

Burns, Inez E. *History of Blount County, Tennessee*. Nashville, Tennessee: Benson Printing Company, 1957.

Dykeman, Wilma. *Highland Homeland, The People of the Great Smokies*. Division of Publications, National Park Service, Department of the Interior, 1978.

Frome, Michael. *Strangers in High Places: The Story of the Great Smoky Mountains*. Knoxville, Tennessee: University of Tennessee Press, 1994.

Hudson, Charles. *Knights of Spain Warriors of The Sun*. Athens, Georgia: University of Georgia Press, 1997.

Kephart, Horace. *Our Southern Highlanders: A Narrative of Adventure in the Southern Appalachians and a Study of Life Among the Mountaineers*. Knoxville, Tennessee: University of Tennessee Press, eighth reprinting, 1998 (originally published by MacMillan Company, 1913.)

Mathes, C. Hodge. *Tall Tales from Old Smoky*. Johnson City, Tennessee: Overmountain Press, 1991.

Myers, Bonnie Trentham. *Best Yet Life and Lore of the Smokies*. Bradenton, Florida: Leadership Options, Inc. 2002.

Parris, John. *These Storied Mountains*. Asheville, North Carolina: Citizen-Times Publishing Company, 1972.

Russell, Gladys Trentham. *Call Me Hillbilly*. Russell Publishing Company, 1974.

Tennyson, Alfred LLoyd. "Ulysses" in *Poems by Alfred Tennyson*. London. Edward Moxon. Originally published October 1883.

Thomas, William Holland. Letter to John Platt, January 25, 1871. Cherokee Museum, Cherokee, North Carolina, presented in the Digital Library of Georgia.

Wilburn, H,C, Great Smokies National Park Historian. Collection of Letters. 1935 – 1936.

MAPS

C. C. Royce map, 1884. Sections on cover and page 107.

"A Map of the Tennassee Government, formerly part of North Carolina taken Chiefly from Surveys by General Daniel Smith," 1973. Section on page xiv.

7.5-Quadrangle maps, Courtesy of U.S. Geological Survey, used for planning. Modified section on page 28.

ADDITIONAL RESEARCH

East Tennessee Historical Society McClung Collection

Great Smoky Mountains National Park Archives

ACKNOWLEDGMENTS

We extend sincere thanks to our families and to all the rangers, maintenance folks, administrative personnel, park neighbors and friends. We will forever treasure your memory.

NOTE TO READERS

The survey markers, blazed trees and monuments mentioned and visually displayed in this book are real. To behold them is an overwhelming experience—a venture into history. To place your hand on these artifacts is to almost hold the hands of the Americans who lived so long ago when our country was new.

Having said this, there is a concern—in fact, a fear we feel behooved to share with you—that these historical gems might not be protected. We hope that anyone reading this book will not attempt to remove or deface these treasures. They, indeed, belong to all of us. Take only pictures, leave only footprints.

ABOUT THE AUTHORS

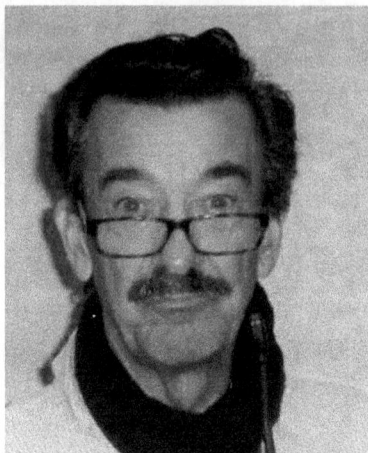

Dwight McCarter

Dwight McCarter was born and raised in the Smoky Mountain area and was employed as a backcountry ranger by the Great Smoky Mountains National Park from 1967 until he retired in 1994. He is a nationally famous tracker who has located numerous lost persons and criminals. He has explored and documented vast and remote regions of the Great Smokies. He is the author with Ronald Schmidt of *Lost, A Ranger's Journal of Search and Rescue* and coauthor with Jeff Wadley of *Mayday! Mayday! Aircraft Crashes in the Great Smoky Mountains National Park 1920 – 2000*.

Joe Kelley served as national park ranger for more than thirty years, mainly in the Great Smokies, and on the Blue Ridge Parkway. Special assignments nationwide have included the National Park Service Special Events Team, search and rescue operations, forest fire management and presentations of interagency training. He and his family lived and worked with other Great Smoky National Park staff and neighbors who were raised in the area before the park was established.

Joe Kelley

www.ingramcontent.com/pod-product-compliance
Lightning Source LLC
Chambersburg PA
CBHW070804280326
41934CB00012B/3047